The Exposed Engineer

**Essential Success Principles for
Engineers and Scientists**

Tomasz D. Jasinski

ISBN: 9798675613175 (paperback)

Cover design: Pixel Studio
Cover graphic: Jegas Ra

Printed in the United States of America

First printing: Sep 2020

Dedicated to my wonderful parents, both engineers,
for starting the journey

Contents

Introduction

After twenty years in engineering I can say that it has been a monumental roller-coaster ride, yet it seems like only yesterday that I started out. I've developed complex electronic systems, written production software, flown in military aircraft, been out to sea, worked in several countries including outback Australia and across the US, hired many talented engineers, led large teams of engineers and scientists, hosted discussion groups at conferences, and worked with engineers and scientists from some of the world's most important scientific and cutting-edge technology establishments. I feel like it is only the beginning, and I feel like I am only now getting a handle on it.

If you are new to engineering, then you are potentially in for an extremely rewarding career. If you absorb the material in this book, then I believe you will quickly experience excitement and success in your career. Whether you are in it for the intellectual stimulation, or to change the world, or to invent, or to publish, or to earn money, or simply to be the best engineer you can be—it is all there for the taking. But once you get going, you may realise that engineering is quite different than what you had imagined. There are subtleties that you will observe over time—what people strive for and how people behave, as well as the problems you face. The engineering and science disciplines will strike as more unique the longer you are a part of them.

While I predominantly address the engineering profession throughout this book, my advice is equally applicable to scientists. As an engineer as well as a scientist who has worked in scientific organisations and universities, I also appreciate the nuances of a career in science. However, my formative experiences were in engineering which amused and inspired me, so this is the focus of this book.

Welcome to the World of Engineering

When I started out in engineering, I thought that the game was all about the technology—that was my own focus and I did not really see the people as special. It was all about creating or learning about the most outstanding ideas and making an impact. Engineering was all about using technology to impact the world. The impact may or may not have improved the world—this really did not matter, so long as it disrupted the world.

Take the example of my engineering peers. Over the years I have seen some crazy inventions from them: a self-triggering air-rifle set up in the backyard for hunting neighbourhood cats; a do-it-yourself electric bike that passes through traffic and gets you to work faster than a car; an active smog mask that filters air in traffic; a drone that scoots through the workplace, before drones were even popular; model rockets, model aircraft; and remote-controlled everything. These are some of the things I have had to deal with in a workplace of engineers—a utopia for clever minds allowed to do what comes naturally to them, which is invent creative yet often disturbing technology.

These are the impressive and often humorous aspects of engineering. There are also the impossible levels of shyness and introversion, the awkward silences, the scalded palms from chemical burns. I have seen people dress and present themselves so badly that it is impossible to think how it was done, and I am not exactly a fashion stylist. I swear that some engineers roll out of bed and come straight to work, their hair strewn in all directions, wearing coffee-stained t-shirts and with beady, tired eyes after successive midnight video game, card game and drinking sessions. Then there are of course the endless detailed discussions of fantastical creatures from board games, movies and beyond.

Engineering is a unique discipline, but the more I get to know it the more I realise it is the engineers that make it unique. It is their attitude towards all-important technology, making cool and disruptive solutions, while maintaining healthy scepticism towards the humanities and social sciences. What on earth do these people do? What do they build or create? Engineers also see through a facade very quickly. We know when someone does not know what they are talking about. We shun pompous behaviour and political games. We take pride in building, developing and fixing. We place a huge importance on learning. Engineering is, after all, rooted in the premise of continuous learning.

If you are already an engineer then I am sure you love your work, knuckling down and solving problems as well as coming up with cool ideas and implementing them. You love learning about new technology and interesting facts and histories of inventions. If you are like me, then you do not enjoy talking too much. You do not expect rewards or recognition. You value a job well done. You shun people with large egos, or people who overdress, trying to impress with mere fabric, or people who use complex technical jargon to describe simple concepts. You can immediately sense these people from a mile away. You can *smell* them. You probably have an unimpressive wardrobe. You admire people who are genuinely good at something and despise the pretenders.

Why was this Book Created?

The longer I have worked, the easier my career seems to have fallen into place. I suppose that once you commit time to your career, you find a rhythm and effective approach, and establish a clear direction. It then becomes easy to contribute work in the right places. Several years ago, I looked back at myself starting out as a graduate, and wondered, "What would I have told myself twenty years earlier?" I had puzzled over this for some time, and over the years have come up with some suggestions which I maintained as a collection of notes. Of course, I regarded this thought as purely academic; even though I would have saved myself many years, how could I possibly travel back in time to speak to myself?

It eventually occurred to me that there are millions of engineers starting out all over the world, in my exact position! On top of that, there is really a lack of valuable, down-to-earth sources that guide young engineers on their journey. There is no shortage of experienced and successful engineers out there, but they normally do not have time to present their best knowledge (and I suspect that for some engineers their best career advice is a closely guarded secret).

The real reason for my motivation has been seeing new starters come and go who were really making some terrible blunders. It is one thing to see a handful of graduates make mistakes, but I was meeting hundreds of graduates and I was seeing the same mistakes! Often it was a problem with their internal thinking or attitude: an expectation for the engineering world to deliver a career for nothing; a belief that hard work had to be minimised at all costs and success should come at the expense of others; an ingrained belief in individuality. Perhaps this kind of thinking was left over from their school years or the influence of their peers. In many cases it resulted in some super-bright graduates quitting engineering for good!

Please do not think I am condescending—like I have achieved it all, or like I know it all. This is far from the truth. I am on my own journey. Each day I am reminded how little I know; each day I am learning something new, whether it is technical, or something about others or myself. I still have grand dreams and ambitions I aspire to: companies I would like to work for; research I wish to conduct; ideas I'd like to patent; books and papers I'd like to write; positions I'd like to hold; money I'd like to make; and even sports cars I'd like to drive! I am still on my path, stumbling and learning every day, just like every other engineer. This book is therefore designed to give you a solid boost based on the knowledge myself and some of my peers have accumulated to date—it does not contain all the answers to a successful career.

I am also not insisting that you follow my path either. This would be pompous and ridiculous since you must find your own path that suits your own talents and dreams. This book is written much more generally than that. You see, *with time I have recognised patterns in the engineering profession that repeat again ... and again ... and again.* That is, there are things that successful engineers seem to do consistently, and things that poor engineers seem to do consistently, and they repeat over and over again. There are behaviours that will, on average, serve you well, and behaviours that will, on average, serve you poorly. This book aims to identify as many behaviours as possible and present them to you in a clear and easy-to-understand fashion, so you do not make the mistakes that your peers are inevitably making. Make no mistake about it, most engineers are making blunders daily that are seriously hindering their achievements and career progression.

Also be aware that this book has been extensively proof-read by many of the people I respect the most in the profession. In fact, in writing it I have tapped into some great engineering minds so I am very confident that the material within this book will give you a tremendous boost, and I suggest that you refer back to it as you progress through your career since it may be challenging to fully comprehend after a first reading. I have also included handy exercises throughout the book that really allow you to self-reflect and absorb the concepts as quickly as possible. These are important, but do not let them dissuade you from reading the book. If you do not have time to complete the exercises, then I suggest reading through the book once, and if you feel like it was valuable then you can always go back and do them then.

Perhaps you are reading this because you are dissatisfied with your job: you feel like your career is going nowhere; you are frustrated by your colleagues and you are about to quit; or you've just lost your job. I feel for you. I have been in dark times in my career, when I thought all was doomed and the future was bleak. The good news is that you have found this book which shows your determination to get your career on track, and to do what you have always dreamed of doing. On top of that, you are going to walk away with priceless information that others do not have—most people whose careers are going well, or even average, will not seek out a book like this, so consider this a big success! If you are reading this book because you want to improve your career, or get going on the right foot, then I salute your proactivity. Most people do not act until a crisis looms.

Once you have read this book and you start to apply the principles, you can relax—you have a lot of time to get it right, and everything that seemed like an insurmountable obstacle will seem like a small hurdle with a very important purpose when you look back on it in the future. You now have an opportunity to absorb everything that is in this book—basically all the best information that I have, including the most valuable information that I have managed to collect from my peers, supervisors, my own experiences and my own reading. You will soon have a massive advantage over those who never had these hardships and you will quickly surpass them. In fact, this is a key message that I will be conveying in this book repeatedly—just about all great successes stem from failures and difficult times. Each failure is a priceless opportunity for growth and action.

The Five Pillars of Engineering Success

If the fundamentals of engineering success could be distilled into five key components, then the ones below would be close to the mark. All engineering successes seem to be built on these foundations, and virtually all successful engineers possess them. Conversely, those struggling in engineering and science will typically neglect one or more of them, and without these foundations in place, all future progress is seriously hampered.

1. A game plan and acting on your plans without fear

Without a career plan in place, your progress is simply based on chance. It is even better if you can articulate it and write it down. It needs to be clear in your mind, but it is not that simple. Your game plan needs to be strongly rooted in your values and aspirations. You need to think deeply and carefully about what those are, and you need to be thinking with an end goal in mind. Your aspirations need to excite you. This book will help you determine for yourself what is important to you.

Once you have a clear plan in place, you must act decisively! Without action, nothing usually happens, and the grander your plans the more action you need. With bigger plans come bigger obstacles, yet most engineers live with some underlying fear that prevents them from overcoming these obstacles and realising their wildest dreams. They doubt themselves, their worth and their values. They never take the action required to achieve their biggest goals.

2. The right attitude

Taking responsibility for absolutely everything, whether good or bad, instils a healthy attitude that sets you up for success. The top engineers do this by instinct. The poor engineers will blame circumstances and other people for their failures. This gives them a convenient excuse not to act. Furthermore, they never walk away with a valuable lesson.

Giving your services without the expectation of a reward is another key trait that is evident in the top engineers. This philosophy turns the common attitude seen in engineering and science upside down. Most engineers seem to be thinking only about themselves, and how they can benefit. True success will only come if your aim becomes to serve others and to create value for others. If you think about the most successful engineers and technology pioneers out there, they all made a positive impact on the world.

Finally, the top engineers are humble and always see themselves as learners. They know that in the scheme of things, they are small and no matter what their specialisation is and how many years they have been working, there is always much more to learn about. Humility leads to better learning, a better state of mind and better relationships with people. Realising that you have a lot to learn ensures that you are hungry for information and seek constant improvement.

3. An appreciation of other people

Successful engineers all realise that other people are the key to any meaningful achievements. Other people are a source of ideas, feedback, motivation, skills and tremendous output.

Virtually all great achievements have been carried out by large teams, even if on the surface it seems like it was an individual effort. Relationships with others should be nurtured, cherished and regarded as the single most important element of success. Whether you are working with someone, selling an idea or leading a team, your ability to build rapport, trust, instil vision and ultimately get the job done all heavily depend on how you treat other people.

4. Well-honed skills (or skills that are being honed)

The bottom line is that your value as an engineer or scientist is determined by the skills that you can bring to the table, regardless of what they may be. Your pay, promotions and rewards are all determined by how much value you add. For this reason, the top engineers are always improving and striving to become better at what they do. They are striving to get the job done to a high level no matter what.

It really does not matter where you are with your skills right now. What matters is realising that it is your skills that are desirable in engineering. The great engineers know that they must be able to do something exceptionally well, using exceptional skills. They appreciate that such skills can take many years to master and so they patiently work away at them.

5. Persistence and dedication (doing it 'just because')

Within any engineering career, there are endless and substantial obstacles. Some of them may be technical engineering problems while others can be related to funding, staff and career. These problems catch many talented engineers off guard, stopping them in their tracks. It is important to plough on, and not lose faith, even if the path ahead is not clear. The outstanding engineers have an inner purpose and drive that always ensures that they plough on no matter what the obstacle.

CHAPTER 1

Do You Really Want to be an Engineer?

Long hours; scant rewards; introverted colleagues; average pay; uninspiring social groups; little recognition by society; and limited opportunities to meet new people. You spent four or five years at university while your school friends seemed to be getting on with their lives. When I think about it now I wonder "what was I thinking?" Was I mad? The truth is, as a seventeen-year-old scanning the pages of a university course guidebook, you don't have a strong basis to make a decision on. As I wander the corridors of my workplace now, I see passers-by looking at their shoelaces, or blank stares focused on some distant technical problem. Lunchtimes are spent playing nerdy card games.

Very quickly you are drawn into a lifestyle of video games, movie outings, geeky drinking games (oh wait, perhaps that's just university!). Your existence becomes secluded, protected from many of the other aspects of the world. Discussions concerning politics, money and women are either non-existent, or at least not undertaken with much enthusiasm like they are not major elements of life. Some aspects of these realities are harsh and indeed you are better off living on this secluded geeky island, but many realities are wonderful and you are truly missing out. There is, of course, a time and place for everything, you tell yourself.

Of course, I am being cynical and slightly facetious here. There is the potential in engineering to fulfil and greatly exceed your wildest desires from your technological achievements, to pay, to job satisfaction, to working with the smartest and most interesting people. But it won't be handed to you on a plate, or come out of a box with a ribbon tied around it. To get there you will have to work hard at it and first of all learn, change and adapt substantially. This is exactly what will happen in every other profession, if you are to become truly successful.

The bottom line is, as with whatever other career path you may take, you must enjoy it, and you must be driven. Whether you are good at it initially, or whether you have an appropriate background or whether you went to the 'right' school really doesn't matter in the long run. Your decision to commence this career should be based on your passion, not your background or what others are telling you. The point of this chapter is therefore to highlight that engineering isn't fundamentally any better or worse than any other career. It just 'is'! It carries its own pros and cons. *The important thing is to want to be an engineer; to have a strong desire for solving problems and working with technology no matter what.*

Regardless of the career path you take, you are only going to succeed, progress and receive recognition and satisfaction if you constantly improve and eventually do it well. You will only do something well if you commit to it and push through all of the inevitable obstacles. And you will only persevere and push through the obstacles if you are passionate and truly believe it is your calling!

Make no doubt about it, there will be times when you will become frustrated, and feel like you are going nowhere. You will be at your wits' end and seriously questioning yourself, your career and the world. If you are not deeply and emotionally committed to what you are doing and why you are doing it, you will quit. I've seen this happen a number of times. Worse still, even if you don't quit engineering, you will settle for life as a frustrated, mediocre engineer, never really fulfilling your potential in any aspect of engineering.

I would not advise getting into technology because it is the hip thing to be doing at the time, or because somebody told you so. You should carefully think about why you are doing it, to determine if it is not a shallow or ego-driven desire. You should feel something deep down in the pit of your stomach telling you that you will continue no matter what! Having said that, it is far better to make a move that may not be entirely right, than make no move at all. Engineering has a reputation as a bit of a default profession. In other words, many of the bright and technically capable friends I had enrolled in engineering because they could not think of anything better to do. This is also fine, I think, since it is a flexible profession and you can adapt later down the track once you figure out more accurately what your passion is and what you are good at.

It is far more important, in my opinion, to pursue what you love from the get-go, rather than commence a career in engineering just because your friends did it, or your dad was an engineer and wants you to continue the family tradition. It only becomes apparent later on that the world is a big place and you have a lot of time to get your career right. Just about anything that you would like to do, you can do, *but you have to be on a path.* My best advice is to trust your heart (for example, you absolutely love making people happy), and not your logic (for example, you heard that there will be a shortage of accountants). If you must be analytical, write down and rank the top ten things that you absolutely love doing (for example, programming, travel, video games and so on) then see which profession satisfies the most. If you can satisfy five criteria, you are doing well. You probably won't appreciate the importance of this simple exercise until much later in your career, so my advice is "do it!" It is a much more valuable exercise than it seems on the surface.

On that note, I have placed simple exercises throughout the book to really get you thinking and absorb the material. Although it feels uncomfortable addressing these issues head on, I believe it is important to get into the habit of doing this type of self-reflection exercise in the long term. It is far better to attempt each exercise in one or two minutes and do a poor job of it, than to not attempt them at all. The sheer act of re-cruiting your brain on these exercises will place in action the machinery that will eventually solve these questions. And the best part is that it will all happen subconsciously.

Generally speaking, when it comes to career planning and progression, naivety is not your friend. I would therefore rec-ommend putting pen to paper, and making a quick attempt to answer the exercises. You don't have to get it right. In fact, the first time you do these you will really struggle. It took me many years and many attempts at these exercises to confidently be able to answer them. Below is the first one.

> Exercise 1: Write down and rank in order of importance to you, the top ten things you would enjoy doing if money was no concern. They don't have to be work related.

In my case, I loved all things technical and developed an early hobby in electronics, but I wasn't particularly good at it nor was my hobby particularly engulfing. However, it was my higher-level and more abstract desires, such as making society better through technology and always striving to build something novel, that have been far more valuable. At times when things are not going well, I just remind myself that I am making the world a better place in terms of technology, and I am on the path to building something novel, and my sanity is automatically restored! This brings me to the next point—figuring out at a deeper level what is driving you from within.

Figuring Out What Drives You

I believe it is of critical importance to have your own set of values, or 'axioms', as I call them. Axioms in mathematics are a set of rules from which all other equations are derived. I believe that the vast majority of people don't have an awareness of their fundamental axioms or values. They may know what they like and dislike, but this is much higher on the ladder of reasoning. For example, they may like a particular programming language, or like a particular kind of person, but these preferences must have a solid and fundamental basis, like the foundations of a house. If the foundations of the house are weak then the whole house is weak, particularly in testing times! I believe it is important to consciously think about what your values are and to become aware of them; clarify in your mind what your values are and why they are important to you. You certainly have values, but if you've never thought about them then it may take some prying to bring them to the surface.

Not understanding your values is fine for the first several years of your career, *but after a while the nature of problems that you encounter become complex, and you desperately seek rules by which to operate.* Sure, there are company policies, and laws, but you will not always find the answer and will seek out some deeper form of guidance. I see a lot of mid-level engineers stumble at such times. If you have a solid set of values, you will always know what the right decision is. Sometimes these values need to be refined. They can naturally change, drift or mould depending on experiences but I usually find that the most fundamental values really don't change much throughout your life. Examples of good axioms or values are:

- I love to help people.
- I don't like to be wasteful.
- Optimism is the key to getting anything done.
- Persistence is the key to success.
- You must be loyal to your employer and your colleagues.

- Honesty is important.
- I value compassion.

To that end, below is the second exercise, designed to get you started thinking about your values. Once again, you do not have to get it right, but I would recommend making an attempt.

Exercise 2: Think about what is important to you at a very fundamental level and write down the five key values that govern your life, whether professionally or personally.

Are You Doing the Right Thing?

First of all, career changes do not necessarily represent failures. Sometimes you truly pick something that's closest to your heart, but after several years you evolve and you know it is time to stop and rethink your career. This is perfectly acceptable and common. This is a result of your development which is never-ending. The only time that progress should stop for you (and your desires remain fixed) is when you are no longer living!

There is a common misconception that you are heading towards *something* that is a final destination (perhaps a steady job and steady pay cheque, a qualification or perhaps a house and kids) but the reality is that you will never feel like you have arrived! This is not because you won't achieve your goals (you probably will if you stick to them) but along the journey new goals will appear, and you will develop new preferences and outlooks; you will mature and change. This is perfectly normal since your career (and your life, for that matter) is a path, and not a single goal or destination. You will always have goals and you will always feel like you are not quite there yet. The important thing is to be on a path and not stalled, scratching your head as to what to do next, or wallowing in fear or frustration. Of course, no-one can tell you whether you are on the right path or not. Be very wary if someone attempts to do this! You can only know this from your heart—from how content, happy and at peace you are. *If you are on the right path, then just doing what it is you do will make you happy, with no need for any destination!*

Every person has their own dreams and goals. Sometimes it takes time to figure these out, and sometimes they change along the way. This is fine. The main thing to remember is what is valuable to yourself, is not necessarily valuable to the next person. For example, one person wishes to set up a business in computing software, another wishes to travel the world and blog for a living, yet someone else may like to work for a large organisation or the government. These are all fine. You should never judge these ambitions according to any monetary system (i.e. how much money you make) or other arbitrary value system. *True happiness comes from pursuing what you love. Misery comes from ignoring what you love and were meant to do, which often is the result of following other people's desires.*

Exercise 3: List three dreams or goals that you love working towards, whether in your professional or personal life. Often these are things that make us lose track of time.

Once you've recognised what you love working towards, then you should take the necessary steps so that you can work towards it as much as possible! I will explain throughout this book how to do this. *And once you are doing what you love, then you should love what you do, despite the ups and downs!* Find an emotionally compelling reason to work hard at it, and success will be merely a formality—almost a mathematical equation. Working away at something with compassion and emotion, and believing that it is truly what you were placed on this world for, will not only bring you success, but it will make your work life easy. Your day will flow effortlessly and you will feel less worn out and open to new ideas and progress in general.

I would also suggest working towards something that is worthy of your time. For example, if you are a mathematics genius then you may also excel at keeping budgets. There is, of course, nothing wrong with keeping budgets or just about anything else you may be good at. However, you are more likely to achieve acclaimed success at things that are harder, rarer, and generally require a bigger investment in resources to succeed in (such as effort, education, natural ability, money and time). When combined with your passion in a particular field, you are far more likely to develop unique and highly sought-after skills excelling at something that very few people have (or are willing to put in) the resources to succeed in.

Also, a final warning—always look beyond your self-assessment of your potential. *Your potential is almost always far greater than you can comprehend at the time.*

Changing Direction

How do you know when it's time to throw in the towel or change direction? I hope that means with your particular company, and not your career! I've met many people who have quit their job (this is obviously quite common) but I have also met a number of people who have moved on to do something outside of engineering. There is nothing wrong with this. In fact, as a generalisation, the most successful people in the world have made numerous, significant changes in their lives to get them where they are today.

The million-dollar question is, of course, when do you move on to something else and when do you stay? Hopefully by this stage you have done the three exercises and you have a reasonable understanding of what you are passionate about and what drives you. The reason this question is so difficult is that there is no right answer! It is highly subjective and highly dependent on you!

Generally speaking, if I find myself day after day feeling like I am wasting my time, and I am feeling unhappy, then I know it is time to move on. Your subconscious mind knows when you are on your path or not, and though it won't tell you in words, you will just get that feeling deep down that this is not right for you.

It may be time to move on when you are regularly thinking to yourself:

- What am I doing here?
- Is this what engineering is really about?
- I am wasting my time in this job.

- How did my life come to this?
- I need to get out of here now!
- I can't stand these people anymore.

These are just typical thoughts that you may experience that indicate it is time to move on. Generally speaking, if you feel out of place, like you are not heading where you want to go, and like you are wasting your time, then this means you need to change something.

Conversely, if you are happy, time is flying, you are busy, you can't wait to get out of bed in the morning, you look forward to getting to work, you are keen on planning your days ahead, you look forward to seeing your colleagues and you feel like your time at work is generally well spent, then perhaps things aren't that bad. Note that just being busy is not an indication in itself that you are on the right path. Plenty of jobs will occupy your time. You must feel as though you are on the path intended for you. You must feel satisfaction, challenge, happiness and an eagerness to continue. Every job will have its frustrations, when everything seems to be stacked against you, but when the eagerness to overcome the problems isn't there, when you have lost the drive, and when overcoming the problems becomes pure misery, then perhaps it is time to change something or move on.

It should also be stressed that sometimes you will outgrow your job and the people you are working with naturally. Six months ago, you may have felt like you had the best job on the planet, but now it bores you. Don't think that this necessarily represents failure on your part. It is natural to outgrow your job since you are continuously growing (or at least you should be!) and only certain workplaces will grow at the same rate and in the same direction as you.

If you are generally happy with your job, then you may like to refine what it is you truly love to do. You may like to develop a deeper understanding of your passions by doing exercises such as the ones in this book. You could expand your mind and your social circles through conferences, journals, books and networking opportunities. The reason that I recommend expending effort when you are actually happy with your job is that the possibilities for improvement are virtually endless but we can rarely see the possibilities initially. By expanding your horizons you may aspire to something much bigger!

If you are not happy with your job, then realise that you should be, and that there is something much better out there. *If you tolerate misery, then all you will get is misery.* No-one will rectify the situation for you, since it is not their responsibility. If you decide to make a change and pursue your dream job, then I can virtually guarantee that with persistence you will eventually get it!

It is also important not to do anything rash. Beware—this is an extremely common mistake. This means, do not burn any bridges and treat everyone respectfully even if you don't expect to see them again. Who knows, you could be going back there sometime, or the next opportunity could well be created by your existing situation. By continuing to do your job professionally, to the best of your ability, and by not offending people and spreading rumours, or trying to reduce workplace morale, you maximise your chances of future success. It is amazing how many people who are dissatisfied with their jobs (which is very common) whinge, spread rumours and attempt to discourage others in the workplace by conveying their own pessimism and problems, which may not be shared by others! This is not the correct way to proceed. Very few jobs out there will be right for you, and it is your responsibility to find the ones that are, by lining them up carefully according to your own skills and ambitions.

Realise that it is quite common for your values to not align with the values of the workplace. No one is necessarily 'at fault' here, even though you may like to blame certain people or certain regulations. Accept this and move on. Also, realise that you do not need justification or approval from anyone to make a change or leave. You do not need to convince anyone else that the workplace is bad for you! Trust your instincts. As I've said, I often see disgruntled employees spreading rumours about a company. Subconsciously, once they gain acceptance from their colleagues, I think this gives them psychological re-inforcement. This is completely unnecessary since no-one likes a whinger, plus you should never, ever be dragging other people down. Thirdly, other people may not see eye to eye with you anyway. They have their own career goals and values which may actually align nicely with those of the company!

Also, in 90% of cases you should not be quitting your job until you have a new one lined up. I mean an actual, official job offer with start date and salary. There are exceptions here, such as when you are moving to another country (you should always attempt to have something else lined up anyway—it can be hard to find something in another country when you can do it in the comfort of your own town with the luxury of time) or you have substantial financial backing to support you while you are looking for something else (perhaps you are taking a gap year or you'd like to try your hand at a small business or perhaps some freelancing). It is always best to assume the worst in this respect, and make sure you have a contingency plan for when you return. In the case of starting a small business, the best advice that I have ever received was to develop the business in your own time while you work your existing job, and only once your company is making a profit large enough to survive on, and you have some savings, and the outlook looks positive, then you can quit your day job and commit fully to the business. Although I tend to be an optimist and value positive thinking, in the case of career changes and new businesses, it always takes longer and it is always harder than you expect. Therefore, make sure to put additional effort into your preparation, your risk mitigation and your safety net. It is best to track down and talk to others who have already done what it is you would like to do.

Generally speaking, I see people get worked up emotionally and then quit their jobs before they have a replacement worked out. Keep actively searching while you are in your old job. Talk to as many people as you can and establish networks. Set up a program of who you will talk to and what applications you will write. Be disciplined. Whether it is a new job, a business, freelancing or a break, develop a plan on paper to the smallest detail before you take the plunge. Keep searching until you have found something that brings you closer to your goals. It is unlikely to be your ultimate dream job from the get-go, but if it excites you then it is likely to be a step closer!

CHAPTER 2

Building Your Foundation

I f you are still here then it means you are committed to hitting the ground running. Or perhaps you have already started your career and now you are determined to learn as much as you can to get where you want to be. Congratulations on taking the first big step and seeking advice!

The first few years in a new workplace can be exciting— there are many new experiences to quickly learn from; you have an appetite to absorb and learn new skills and you are motivated to attack tasks with vigour. However, sometimes the first few years may not be what you expected. The work that you are doing may not line up with what you were taught at university and you may become discouraged. You may feel like you are out of place and you have made a mistake with your career, or that your supervisor is from a different planet and not in touch with you. You may feel bored, reading endless documentation designed to bring you 'up to speed'. These are some of the most common stories I hear. The harsh reality is that there is no 'normal' start and the views and priorities of your employer may be drastically misaligned with yours, but this is not something that should worry you in the early days. Your employer has their own ideas of where you fit in and where your future lies. It is normal to feel out of place initially (after all, unless you landed in academia, your first job will be very different to academic life), but it is important to carry on and do the best job that you can. In the early stages, *it is important to absorb as many new experiences as possible.*

Your Schooling Up Until Now Means Nothing

The harsh reality is that the wonderful grades you got at university (or those terrible grades!), as well as all those awards you got in high school, and your deep belief that you are exceptionally smart, or gifted, or privileged to have made such a wonderful start to adult life, means nothing now. These are as relevant as your preschool report card. You are now starting from scratch. You may not realise this now, but your true value is hidden inside you. It is your attitude to problems—the way you handle people, your drive, the language that you use, your body language, your level of motivation and your eagerness to learn quickly. Those subjects and grades were just a test to get you here! This is hard for young people to grasp, since they think it is their grades that are of value, and that their grades define them. This 'reset' will happen again and again throughout your career. For example, if you land a management role, you will become a beginner once more, this time to the challenges of managing people. *An engineer's life is all about constant change and learning.*

As harsh as this sounds, *the engineering world owes you nothing because of your grades.* You won't necessarily be given an interesting project, or a big pay packet, or work with nice people, or have a nice boss. Consider it a bonus if you do receive any of those things. Such privileges must be worked towards. I often hear new starters announce that their schooling entitles them to an interesting career, but this is not the case. Sometimes I come across the attitude: "Okay, I have graduated, now what does engineering have in store for me?" We are too used to having tasks and guidelines hand-delivered to us throughout school, and we maintain this expectation throughout our early working lives. The problem is that once we eventually learn that this is no longer the case, we get discouraged, frustrated and pull back, never having the same level of motivation as when we first started. This is a shame, since I see so many bright people lose their way, frustrated by their supervisors or peers, or their work, rapidly losing their confidence and enthusiasm. Rather than accepting that it is a temporary setback, and temporary setbacks will constantly appear and must be dealt with, they believe that it is a terminal misfortune or the way that engineering inherently is. *The good news is that everything that you have ever wanted from engineering is all there waiting, but in most cases it will take time and effort to get it!*

Develop a Career Path

One of the most important habits you can develop in order to ensure that your engineering career remains on track is to make a plan of where you want to go, and update the plan again and again. Sadly, this type of long-term planning is very uncommon, since it requires an honest appraisal and critical thinking, which is uncomfortable. Some would say that career planning removes some of the pleasant surprises from your career, but this thinking is flawed. There will still be plenty of surprises—I guarantee it. However, to get anywhere meaningful, your mind needs a clear target.

No matter where you are now, realise that where you are and where you could be are two totally different places! With the right planning, these can converge very quickly.

In the first few years of your career, while you are gathering priceless experience, this is probably of less importance. However, as options open up to you, you will struggle to know which path to take unless you have a career plan in place.

There is, of course, a type of approach and personality ingrained in some of us which demands that we live out our careers with no planning whatsoever. If this is your style and you can't work any other way, then go ahead and take your career one day at a time. However, keep in mind that you will never feel like you have achieved 'it' until you have defined what 'it' is.

Also, don't just expect your career to fall into place automatically. There will be times when the sailing is smooth. You may have an excellent supervisor that has you moving in the right direction, or your company may have clear goals. You may not realise it at the time, but it is this careful planning of others (just like when you were at school and the progress seemed automatic) that is keeping you moving in some direction. It is highly likely that there will be a time when this path ends or begins to unexpectedly veer off course, or worse still, starts to veer all over the place. At this time, it will be up to you to make your own path. In fact, the further you get along in your career, the more options appear and it becomes up to you to lay this path! There will be times when you are asking yourself "what now?"

Many new starters that I have talked to who are against the idea of making a career plan state that there are too many uncertainties ahead for accurate planning. This of course is partially true. The important thing to keep in mind is that plans can (and should) change. Life is a journey. You don't know where you are going to end up, but that is normal and shouldn't dissuade you from establishing firm direction! This advice may be bitter tasting to the purist, but anyone who has achieved anything big in their careers has achieved it through a staggered path, full of twists and dead-ends, but with strong direction at every point! It is unrealistic to expect a straight path. It is normal, in fact, essential, that there are twists! Therefore, don't be dismayed when sitting down to write your plan by the uncertainty that lies ahead. It may feel uncomfortable, but I recommend that you force yourself and spend at least ten minutes initially developing a career plan.

It is important when putting together a career plan to plan right up until the success that you are after. Don't just plan up to one milestone and then hope to continue with your planning later on. If you would like to work with NASA on launching a space probe, or if you would like to work in renewable energy designing a more efficient solar cell, or you would like to build the fastest electric car on the planet, then this is your end goal, so set up a plan that gets you there, even if the path and the steps are unclear at this stage. It is the end goal that matters most and not the path that you take. The path will form gradually, autonomously and often much to your surprise, and usually in a way beyond your control.

Just like looking out of a boat towards a horizon on the water, you can only ever see out to your career horizon, and no further. There is too much uncertainty to see beyond. As you progress along your path towards this horizon, new options will open up, giving you an opportunity to form new paths towards your goals.

Your goals may change also, and they regularly will! As new opportunities open up, your goals will change naturally. Also don't forget that you are growing every minute of every day. Your brain is not in a static state, but it is constantly learning and readjusting its priorities. You should expect and allow your goals to change.

Exercise 4: List five goals that you would like to achieve by the end of your career and describe in a separate column why you would like to achieve them. The reasons should be emotionally compelling and close to your heart, so think about them carefully.

You will not achieve anything beyond what you have planned. This is an extremely important and life-changing realisation to keep in mind, and something worth rereading! Our early years throughout our upbringing are damaging to the way we progress through our careers. Since we are actually

rewarded for following rules, and being complacent and sub-
missive, it is partly our upbringing and partly our schooling
that encourages us to be passive observers of our careers (even
though many of us can be very hard-working and can achieve
massive success just through hard work alone). I'm not saying
we need to become disrespectful, ignorant and forceful. I'm
talking about being proactive and taking responsibility for absolutely everything to do with our careers. That is, possessing
the realisation that if anything is to be done, it must be done by
you and not be left to others to initiate!

Ideally, you have to know what you want … but be flexible
in your approach. You may not realise it early in your career
while you are still absorbing new information (and hopefully
being guided by your company), but after five to ten years of
professional life this will be of critical importance. Otherwise
you will risk stalling, plateauing and stagnating, which is the
last thing you want in your engineering career.

Dabbling

Closely related to the concept of forming a career path is striking the balance between dabbling in new ideas and committing to known ideas. Some people tend to experiment with new
tools, projects and hobbies all the time, while others will commit to one thing and do it very well. Different people seem to be
predisposed to either of the two, so if they dabble at work, they
tend to dabble at home, and if they like to commit themselves
at work, they will also tend to bite off projects at home that
require several years of attention.

To be outstanding at anything, you have to commit and persevere with it, despite the hurdles that you encounter and despite your urges to quit. If you were to quit whenever the 'going got tough' then you would never develop outstanding skills at anything. When you are learning something new, throughout the initial phases you learn quickly and make rapid progress. Later down the track, once you learn some skills, progress is much slower and the task becomes much more testing on your ability to persevere. It is an obvious statement, but becoming exceptional at something is much, much harder than just getting good at it. I believe that dabblers are discouraged by the difficult phase, which comes after they have made the easy gains. However, this is the most important phase. This is what gives you your value as an engineer. No-one wants to hire a mediocre engine designer. They want to hire the best!

Having said that, early in your career when you are not sure what to focus on, you need more experiences. In this case, I agree with the dabbler's approach—to broadly and indiscriminately learn as many new relevant or interesting skills as you can. Although in this case, a far more efficient approach is to leverage off others, but I will get into this later in the book.

The best philosophy that I have come across related to depth of expertise is the 'comb' model of knowledge accumulation. That is, you pursue some areas to extreme depth (this is what you are employed to do), and you maintain basic skills in many other areas so that you are a well-rounded engineer. In reality, your skills will be a combination of everything in between, but I think that this is a good aim point to have. For example, you may be exceptional at three to five things, but have some knowledge in another hundred areas.

So What is Failure?

There is only one true failure in engineering, and that is when you give up! It may seem like you are limited by technology, or funding, or the rain, but in the vast majority of cases *failure is internally triggered.* This is another one of the lines that will be shocking to you the first time you hear it, so it is worth rereading. The reason behind this is that we normally tend to overestimate where the true limitations lie. They are rarely where we think they lie and we rarely push the limits to find them. This is true with technology but it is especially true when it comes to people.

Secondly, we are accustomed to finding and using excuses in order to set some artificial boundaries, which may not actually exist! Because we put the failure down to forces beyond our control (which rarely is the case), it gives us an 'easy out' accompanied by an inner peace that reassures us that the failure was not our fault.

Realise that you will only fail when you give up and tell yourself it is over. We may tell ourselves this based on previous experiences, or based on other people telling us that the task can't be done. In any case, when we resign to failure, whether personally or as a team or company, that is the only time we truly fail.

Perhaps the only other type of failure, in my opinion, is as a result of no planning or setting of goals. Even though technically by not having a goal you cannot fail, I believe that we often fail to set goals (or don't set ambitious goals) in order to protect ourselves from failure! *Therefore, if you are not moving forward in your engineering career due to lack of a plan, then I would consider this another form of failure. In short, don't make excuses and have an open mind, have a plan in place and think big!*

Time is Precious

One of the worst enemies in your professional life, and one which you generally don't appreciate until later on in your career, is time. When you are starting out, it feels like time is unlimited. Sometimes you seem to have so much time that you have no idea what to do with it! In fact, there seems to be so much time available that you wish you could fast forward into the future. You wish that time would pass more quickly and you could be somewhere else in the future—in a more senior position or in a more established project. Then towards the middle of your career, you begin to appreciate time as a more valuable resource. You realise that you can only carry out a small fraction of the tasks that you would like to. Your skills and personal connections are no longer the main limitation. Perhaps other parts of your life start to impinge on your work life, like hobbies or your family life, for instance. This will further highlight to you that there are only a limited number of hours in the week. Then, towards the end of your career, if you have done everything right like followed your dreams and taken sensible paths, you will crave even more time. You will know exactly what you have to do, and time will become your most valuable resource, ahead of money, friends and nice cars!

Therefore, get into the habit of valuing your time immensely by spending it on quality projects, even early in your career. Spend it on tasks that are important to you, or tasks that will help you get where you want to go (this also highlights the importance of a clear plan). By all means do the tasks that are fun and absorbing, but focus on the tasks that are getting you somewhere, not the tasks that are just fun. Of course, I am not talking about your relaxation time, which is essential. This is something very different. You should definitely allow for time to relax, and loaf around aimlessly. This will, in fact, assist you with your problems at work, since often the best ideas come when you are thinking about other things and you are in a relaxed state.

I've often heard and considered the hypothetical question: "If you knew you had one day to live, what would you do?" Of course, very few people would stick around at work! However, keeping this in mind as a thought experiment is useful, since it highlights the shortness of your professional life, the urgency of the tasks at hand and the need for clearly defined direction at all times.

Prioritise

The normal way of prioritising is according to importance and urgency, but that is not what I'm talking about here because most people know how to do this! Obviously you are going to put more effort into those tasks that matter more, and less effort into those that have little impact. The same applies to those tasks that are urgent versus those that are not.

However, all of your problems can be generally broken up into the problems that you can control and the problems you can't control. A surprising number of problems or tasks lie at both extremes. For example, at one extreme you have complete control over how you talk to your colleagues, and on the other extreme you have no control whatsoever over the weather. There is also some grey scale in between, where you have some control, but generally speaking these represent the two major types of problems. *Most successful engineers place a disproportionate amount of effort and worry on the things they can control, and no concern whatsoever over the things they can't.* Most unsuccessful engineers do the very opposite—they seem to dwell over the things that are out of their control (such as internal decision making, policies, the attitudes of other people and so on), and not pay much attention to the things they can control (such as the quality of their work, the amount of training they do, the way they interact with their colleagues and so on).

The problem with worrying about everything is that your brain can only handle so much. When you start prioritising problems that you have no control over, then you become seriously side-tracked. I see this very often. Be on the guard for others who do this. Now that you are aware of this, you will spot it quite easily and you will see how days and entire projects are wasted on the wrong problems!

My only other comment here is that the old adage of going where it's uncomfortable holds true for engineering. Often we brush aside the tasks that seem uncomfortable or those we fear.

They seem messy and unpleasant, and often carry the highest uncertainty. This is often an indication that they should be done first! Don't put aside those tasks that you fear. Tackle them head on through discipline. You will be surprised how easy the vast majority will be once they have your full attention. They will also give you momentum to then tackle even more pleasant tasks. Of course, I am not saying to only search out the difficult jobs and always go where it's uncomfortable—this would not be productive! I'm saying that if there is an unpleasant task in the way of what must be done, don't put it aside for later. It will only get harder and it will ride on your conscience, using up the limited resources of your brain. When your first instinct is to put it off, then this is often a sign that it should be done right now!

Making Decisions

I find it surprising how commonly young engineers have trouble making decisions and sticking to their decisions, given that engineering is all about problem solving. Of course, I was no different and I still have to force myself to make decisions sometimes, but I am quicker than I used to be, and perhaps that is why I can sense the difference. Engineers hate to make mistakes, yet it is through mistakes that learning occurs.

Rapid decision making also encourages you to take responsibility for your actions, as well as develop leadership qualities. Your entire professional life and how you navigate it will be determined by the decisions you make, and like all things, practising regularly will make you skilful. If you are a natural at making decisions quickly (you are decisive), then you can skip to the end of this section, but if you are like most young engineers, then read on!

It is also a well-known observation that successful people, generally, tend to make decisions quickly, but they are slow to change their minds. This initially seems counterintuitive to the young engineer. So if you would like to pursue a topic, a technology, a hobby, and something draws you in (hopefully something initiated internally from your heart, and not just a passing fad) then give it a proper go before switching to something else. Become good at it. Trust those instincts that drew you to this persuasion in the first place. It is remarkably easy to doubt yourself, and move on to something else. This is exactly what unsuccessful people tend to do. Unsuccessful people, generally, take a long time to make up their minds to do something, and once they do it they are very quick to give it up. For example, they may spend forever determining what new skill or hobby they wish to learn. They talk to their friends and read everything they can. Then finally, after many weeks or months, they make their decision, but once they start learning this new skill they very quickly decide that it is not right for them and give it up, resuming their search. Often it is that characteristic of spontaneity that draws you to something, without a solid, rational reason, that will result in you having the most success. Often you'll make these discoveries 'by accident'. It will seem like a complete coincidence that you have made a beginning in this new direction. For example, you caught up with an old friend who made a random suggestion. Or you met someone who gave you a great idea. Or you were in a cafe, and stumbled upon a magazine with an ad for a gizmo or training event. Or perhaps you had an insightful dream. These events seem fortuitous. The best career turns and new hobbies that I have undertaken have appeared out of nowhere and were based on what appeared as pure luck. They were not conjured up, or engineered to meet some market need, or the needs of my friends or peers.

Conversely, it is often the analytical approach that brings around ideas that are less likely to stick. For example, you may assess the market, assess the clients or universities, put together a spreadsheet, and make a theoretical determination of what to go for. This approach has its place in engineering and business of course, but it is not necessarily effective for picking a career direction. In fact, taking this approach is not only ineffective for this purpose, but it can be damaging since it can 'mask' you from seeing reality as it is. For example, let's say that you theoretically determine never to pursue involvement in the mining sector. This may block you from any subsequent positive opportunities related to mining.

To summarise decision making in a nutshell (and this is particularly true when it comes to determining your goals, career path and hobbies), trust your heart and instincts. The only regrets that I have in my career (and in my personal life, for that matter) are cases where I have not followed my heart because it seemed out of line with the theoretical, rational line of thinking. Ignoring your heart will always result in a missed opportunity and future regret. Of course, there is a place in engineering for analytical decisions that are time consuming and measured. However, the vast majority of decisions don't require this approach. *Make your decisions quickly and stick to them. Practise doing this regularly on the little things (for example, picking an ice cream flavour or a meal from a menu), and it will become automatic on the big things in life and lead to you developing leadership qualities.*

Grind it Out

Anything of value takes immense work and significant time to achieve. Get used to being uncomfortable and grinding tasks out until they are done. As I've said earlier, take the uncomfortable path head on—don't run away! For example, perhaps some documentation task will require weeks of work and you were hoping to finish it today. Perhaps you found out that you need to ring around for some parts, or find the appropriate experts in the field. Perhaps you realised that you have to learn something new, or find a company that does something unique and you have no idea where to start! Perhaps you have to spend long nights working on a problem, or you have to travel across town to meet with a supplier. What a pain! This is exactly what stops most people from making progress—particularly young engineers. They do not like to tread where it is uncomfortable. They haven't yet learned that being uncomfortable is part of engineering. If there was one lesson that I could convey to young engineers, which I consistently see as the difference between the experienced professionals and the inexperienced engineers, is that the young engineers do not go where it's uncomfortable. Most of them procrastinate, make excuses and put things in the 'too hard basket' way too quickly. As an engineer, you must constantly display the tenacity of a Rottweiler when it comes to attacking problems!

As I've said before, not fearing being uncomfortable does not mean that you should go out and find an uncomfortable path, and attack it just because it is uncomfortable! That would not be an effective use of your time! However, you should not be discouraged if what you would like to do appears difficult. Great achievements are rarely easy, and the fact that the path is hard may be an indicator that you are on the right path to significant success. Having said that, the hard problems are rarely ever as difficult as they initially seem. Our minds tend to go into panic mode and project our worst fears to create the most pessimistic scenario possible. Once you are done with what initially seemed like an impossible task, it will seem trivial in hindsight. New doors will open and you will receive a massive confidence boost.

Exercise 5: List three tasks that you have been dreading to do, but ones that you know must be done. They can be small tasks. Make a deadline for their completion (and stick to it!).

Look at hardships differently to the way others see them. Without hardships, you get nowhere and don't grow. I have found that every time I have made true, groundbreaking progress there was some hardship involved that initially seemed impossible to solve. Often it was demoralising to the core. Thoughts of changing jobs came to mind, and my confidence in my abilities and understanding was seriously challenged. In other words, I almost gave up in all cases. You can't let hardships rattle you.

Look at hardships as opportunities. You have done something right to have earned the right to experience the hardship. You can either give up, bypass it and hand it to someone else to solve for you, or you can tackle it. I have done both. Bypassing it does nothing for you. It gets you no closer to any goals and won't open any doors, boost your confidence or boost others' confidence in you. If you decide to tackle it head on, and succeed, you will be making solid gains. New doors will open. Your experience and confidence will increase. But be warned, tackling hardships opens doors to new hardships. The better you become at answering questions, the more questions will be asked of you. Don't think for one minute that after you tackle the first hardship, it will be smooth sailing from then on. It may be easy for a while (and you'll surely be on a wave of joy) but eventually you'll hit another hurdle. With each hurdle that you tackle you will become stronger and you will get closer to your goals. You will grow in confidence, your skills will improve, you will feel on top of the world and new doors will continue to open.

Learn to Celebrate Your Achievements

When you finally overcome an obstacle or hardship, even if it is a small one, get into the habit of congratulating yourself. Reward yourself to signal to yourself that you are on the right path. Do this even after the minor accomplishments. Too many people worry about solving their problems, but when they make solid ground they don't celebrate the progress. This may sound ridiculous, but it has an important psychological function.

Congratulating and rewarding yourself will subconsciously bolster your confidence, increase your motivation and keep you on course to achieving more. As crazy as it sounds, your subconscious mind needs to know that it has done a good job so that further successes can follow. Why do you think those professional sportsmen and sportswomen are constantly pumping their fists and celebrating after each point? Of course, I'm not saying you should be pumping fists and shouting "yes!" at the top of your lungs in your office, but you definitely should have some reward ritual set aside for all of your little milestones.

Exercise 6: Write down three things that you have achieved over the past week that you are proud of, and have brought you closer to your goals. Think about these achievements, feel how proud you are, smile and congratulate yourself! Reward yourself. Grab a coffee to ponder or have an ice cream, or a beer or whatever rewards you, or take a quick break and go for a walk. Do something that signals to your body that you have done a job well. Get into the habit of writing down and celebrating your successes at the end of each day.

Receiving Your Reward

When it comes to Receiving Your Reward, be aware that there is a natural and universal reward mechanism at play which also applies to engineering. When you work hard on a project and do most things right, eventually you will receive the spoils. The twist is that it will usually happen at an unexpected time, in an unexpected way. Your reward will rarely come in the form that you were expecting. Realise that this is normal. Many young engineers working towards a promotion or a particular position are frustrated that all of their hard work is not getting them over the line. Some of them become sceptical, bitter and eventually lose hope. However, working towards something will sometimes reward you in other ways. Your efforts may be noticed by another employer, or throughout your streak of determination you find a new client. In almost all cases, what you get is actually better than what you had hoped to get, but you must continue to work towards that original goal.

The important thing to remember is that you must have faith that your hard work will be rewarded. It will! There is a natural order to the universe that says it will. However, your ultimate reward is almost always over the horizon. You can't see it, and you can't always see a path towards it, and rarely can you see all of the obstacles. Worse still, often you don't receive any feedback to say that you are getting closer (though sometimes you will, which I discuss later in this chapter). *It is therefore important never to lose faith that your efforts will be rewarded.*

Discipline and Improve Yourself

Just about anyone can do what they enjoy, when they enjoy doing it. It is much harder to do what you know must be done when you don't feel like doing it! I'm sure that you have found that some of the most important work that you have done has happened because you were forced to do it by some external influence or event. Normally, you would not have had the discipline to push on alone. For example, take university exams. There is no way that the vast majority of students would have the discipline to stick to four or five years of engineering study if not for the threat of failed exams! You studied and crammed because you had to! At work, it is usually your boss or sometimes a mentor who absolutely insists that you have to get the work done, preventing any possibility of procrastination.

Progress can also be made because of internal drive. We can all put up with some level of procrastination. However, sometimes a point is reached where the ramifications are just too big and we have to take action. It helps if you put yourself in a situation where the work must be done, such as making a commitment. The commitment must be addressing a genuine need—otherwise we would just quit! You have to establish a good reason for doing it up front. Then it takes discipline to stick to the task and form a habit. We all know from going to the gym or practising a musical instrument that once you get over that first month or so and establish a habit, the work just seems to happen by itself with minimal pain. It also helps if you break the work down into smaller pieces and plan to do the work up front. And be sure to have a final vision in mind—a destination.

Throughout your engineering journey, you must improve or you will stay in the same position, on the same pay grade, limited to the same type of work. I can't overemphasise how many wonderful and captivating engineering projects are out there, waiting to be tackled by you! You can accelerate the process through self-improvement, where you take proactive action to improve. You can also achieve improvement through formal training of course, but it is the mentality of self-improvement through self-driven and continual ongoing improvement that is really what I'm talking about here. You take responsibility in improving in tiny little chunks, every single day.

In a nutshell, try to be slightly better at something compared to a day before. If you are writing a report, then use a few new words, write faster, or write clearer. Exceed your boundaries so that the next day you feel like you are slightly better qualified in what it is you are doing. Whether you are coding, designing, discussing or negotiating, focus on getting slightly better each day. Ask yourself: What would it take for me to improve in this? Who can I talk to? What book can I read? What new challenge can I face?

Self-Esteem is the Key

I cannot overemphasise the importance of a healthy self-esteem. Of course, we have all experienced over-confident and arrogant people, who think they know it all (but often don't), so it may be of some surprise that it is important in engineering. Self-esteem is particularly important in engineering, as many engineers are natural introverts, and tend to undersell their work and underestimate their abilities. All new starters, whether engineers or not, have a tendency to worry about how they are perceived. This is very common. I would say that the vast majority of new starters live in a shell—they are scared to come out, fearing they may offend someone or embarrass themselves. They do not sell their ideas with motivation since they are afraid of how they may be perceived. They don't realise that as long as they remain respectful and professional, a belief in oneself and one's abilities will only make them shine and stand them out from the crowd. If you don't think this is a common problem, then ask yourself: When was the last time that an engineer was able to sell you anything?

A strong self-esteem gives you the confidence to chase your goals—to maintain faith, to push through all of the barriers. It is an important ingredient, and when combined with the other traits outlined in this book, it will get you beyond your wildest desires. Note also that, from my experience, it is much harder starting out with a low self-esteem and having to build one, than starting out with one that is too high and having to curtail it.

You can practise improving your self-esteem by getting your ideas out there—by making suggestions, selling your ideas and implementing them. It really doesn't matter if you make mistakes (so long as you are not making the kind of mistakes that will lose you your job!). This is perfectly fine and will be expected by your peers and supervisors in your first few years (or even decades—in fact, your entire career should be riddled with mistakes and this is normal. I explain this further in the subsequent chapter). Have your voice heard. With your successes, your confidence will skyrocket and you will form a positive inner reinforcement feedback loop.

Modify Your Approach as Required

Throughout this book so far, I have stressed the importance of having a plan. However, it is very unlikely that your career or any part of your work will match this plan precisely. In particular, it is the inevitable obstacles that will derail you. As a result, it is important to be able to think on your feet and modify your approach as required. For example, if you are working with a particular piece of software, and it becomes blatantly apparent that no matter how hard you try, it will not do the job, then you must make a change. Try different software, try a new approach, or make some other change. Persevering with the same software would be like banging your head against the wall. Be creative and think outside the box. Talk to others and read the relevant journals and industry magazines. It is also important to be able to make the distinction between knowing when to persevere, and when to modify your approach—don't throw out all of your good work at the first hurdle! If you are not sure what to do, don't be afraid to talk to your supervisor and others in your field for a second opinion.

You may not realise this, but the successful engineers are constantly modifying their approaches, sometimes many times per day, until they hit upon something that works! Don't be afraid to change direction. Remember, your ultimate goal and what your company wants out of you is to get the job done. How you go about it and how many changes in direction you make really doesn't matter in the end. This can be a problem when young engineers' egos are connected with a particular direction, technology or outcome and they are reluctant to give it up (I talk about the ego in detail in Chapter 4). Don't fear that changes in direction will represent failure, or will give the impression that you do not know what you are doing. The most high-performing engineers that I know (particularly the new starters) are constantly experimenting with new ideas, and experiencing mini-failures over and over again. This is what makes them good! They seem to bounce from one solution to another. This is very reassuring to me as a manager because when I see this behaviour I know the problem will eventually be solved! The same engineers are normally the ones who exhibit that highly sought-after dogged quality, which brings me on to the topic of perseverance.

The Importance of Perseverance

Perseverance is the key ingredient in getting anything meaningful done in your engineering career. There is no doubt that the grander the goal, the greater the difficulty in achieving it. The grander the goal, the more obstacles you will encounter. The obstacles will generally be unpredictable and they will be never-ending. They cannot diminish your drive or your self-confidence or you will not overcome them. You must realise that they are part of normal engineering life. Therefore you must accept them and tackle them one by one. Once you truly accept that obstacles are normal, it feels like a huge weight has been lifted off your shoulders. Once you realise that they don't represent failure on your part, or that the engineering world isn't out to get you, then you can focus on solving them.

Obstacles also tend to be designed with a dark sense of humour. They are effectively questions that are being asked of you, one after another, such as "what are you going to do now?" Sometimes the obstacles that I have faced can only be described as comical. It is as if they were written for a theatre show. It is at these times that you can only laugh and appreciate the sense of humour of the engineering world. Perseverance (and a sense of humour in many cases) is absolutely necessary to tackle successive obstacles. This is one reason why you must always have a specific goal and a compelling reason for achieving it. Otherwise, the relentless obstacles will have you questioning the task, yourself and even your career, and you may eventually give up or not give it your best shot.

The other pattern that I have observed time after time is that sometimes your grandest goals are just beyond the horizon when you hit a massive obstacle. You can't see them yet, and you can't see a path towards them. It is often the biggest obstacle that strikes right at the end. It's as if the greatest test has been saved to the end. Once you tackle this final, disheartening obstacle, there is smooth sailing ahead—at least for a while. This has happened to me countless times. I invested massive effort in getting a project to a certain point, assuming I would succeed. After overcoming a thousand obstacles, something which seems impossible to overcome appears right at the very end. A showstopper. These disheartening 'showstoppers' drain you of your enthusiasm, and they always have that failure stench about them. The vast majority of engineers when faced with such obstacles give up! They say things like:

- "Oh, we couldn't get funding" or
- "The boss won't let us have the money" or
- "It is against company policy" or
- "Those work, health and safety rules make this impossible!" or
- "That guy is impossible to work with. We can't succeed with him on the team".

And that is where the project remains, whereas this is exactly when you need to persevere—at times when it all seems impossible and there is no solution in sight. Now you know that this often means that you are just an inch away from success!

Just to give you an example of this—recently I organised some measurements of a data-link across a waterway. It took around six months of effort to organise the measurements. The team prepared the hardware, we tested in the lab, we overcame the numerous certification issues, we held meetings every week, we battled for funding, we contacted local councils for permission and we put together the right team of people. The gulf was thousands of miles away, so we flew out there and took detailed measurements. We flew back and resumed our planning. Finally we had a day in mind when we would ship tonnes of equipment out and make the measurements.

It was a sunny Saturday morning and all was in place. Then, out of nowhere, a 9000-tonne cargo vessel came into direct line of sight, dropped its anchors and proceeded to start unloading its cargo of wheat. The vessel was blocking all chances of getting any measurements done! The odds of it parking there must have been a thousand to one! There was no budging it! It was due to remain for over a week while we only had four days for our measurements. The engineering world indeed had a sense of humour, but our measurements were not to be! Unfortunately, we would be going home with our tails between our legs. The captain was nowhere to be found, and the ship was rightfully there according to a schedule that we had overlooked. There was no way out. We would have to reschedule.

Fortunately for us, in another strange twist, in my diary of contacts I found the name of a wharf controller that I had written down by pure chance. We called him to explain the situation and within thirty minutes, the ship was gone! We could not believe our eyes!

Following that obstacle, we proceeded to carry out one of the most successful tests that I can recall. Of course, the vast majority of teams would have given up, conducting their tests at a different time or in a different location, but we persevered and were rewarded. For us, giving up was not an option, and the solution wasn't that difficult. The biggest and most disheartening obstacle came at the very end, just before the biggest reward!

Learn to Accept Failure

Not only will there be plenty of obstacles, and some of them will seem insurmountable, but you will have plenty of failures. You will be hit again, and again, and again (do you still want to be an engineer?). Yes, as an engineer you will fail repeatedly.

This is why it is important not to see failure as 'failure' in the traditional sense, but as a necessary learning experience or temporary hurdle. This may be hard to come to grips with for young graduates, who have learned to define failure as something negative, perhaps through the assessment process at school. In this case, everyone wants full marks. And those who perform poorly rarely walk away with a valuable lesson.

Engineering in real life is slightly different. The majority of failures are positive, or at least they contain a positive lesson as a silver lining. In fact, they are essential for any progress to be made (of course, I am not talking about catastrophic failures such as collapsing bridges, though these sorts of failures probably carry worthwhile lessons as well). For every major success I have experienced, I have experienced a hundred failures. For every successful sales pitch, there have been fifty unsuccessful ones. For every successful job application, there have been forty unsuccessful ones. For every successful code compilation, there have been ten failed ones. For every efficient algorithm, there have been ten inefficient ones. Are you getting the picture yet?

It may seem disheartening to the purist or to the theoretical mathematician or physicist. Unfortunately, most facets of engineering are complex and based around some element of trial and error. Furthermore, becoming great at anything requires plenty of mistakes. No successful engineer will tell you that their career has been mostly free from failure. Therefore, paradoxically, if you are not regularly failing, then you are doing something wrong! You are not pushing the boundaries and you are probably not going to outgrow your current abilities (or at least not quickly). This is, of course, fine if you do not want to make any further progress throughout your career.

The old saying that practice makes perfect applies very much to engineering. Not only is it true where you are trying to brush up your skills, whether they are technical, presentation skills, or management skills, but by failing at various tasks you get much better prepared to tackle similar situations when they rear their heads again. Failing hurts, but appreciate that everything that happens to you is a learning experience and happens to you for a reason.

Learn to accept other people's failures too. Your hardships, whether through your own fault or because of other people's mistakes, are really blessings in disguise. For this reason, be tolerant of other people's failures. They are each on their own learning journey just like you.

Of course, some level of failure can obviously cost you immensely—don't misunderstand me here. Do not try to fail just so that you can learn a lesson—it doesn't work that way. Secondly, if you don't walk away with a worthwhile lesson, then don't expect the failure to help you. Often I have found that if I haven't learned from my initial mistake, then I have made the same mistake again and again until I finally learned what I did wrong.

Without failures, you would only be working at 10% capacity. In order to expand, we need to learn from our mistakes and we need to be primed into action by our failures. This certainly applies to the failures of others that you witness too. I've seen my fair share of technical mistakes, poor decision making and poor staff supervision. But as a result, I have usually walked away with a valuable lesson as well as the confidence to know that I can do better! The essence of failure is: failure is positive!

Get into the habit of thinking that failures are happening to you for a reason. The highest-performing new starters that I know are willing to push the boundaries and they are constantly bouncing from one approach to another, learning valuable lessons from their mistakes.

Don't be Afraid to Ask Questions

The old adage of 'don't be afraid to ask questions' really does apply to engineering. Be curious and inquisitive. Ask those questions that you fear may make you look stupid—practise doing so! It is more than likely that the very question that you are pondering over is being pondered by others on your team. If you have faith in your abilities and possess a reasonable amount of self-confidence, then you should not be afraid of asking questions. Be particularly active asking questions early in your career while you have the opportunity. Later in your career, when you become the expert, it will be increasingly difficult to find people experienced enough to answer them.

Asking questions gets you exactly the information that you need, immediately. The answers you receive open the door to further curiosity and contemplation, and lead to a higher level of understanding. The smartest people that I have worked with have had the ability to ask succinct, well-placed questions, which in many cases I had been pondering myself! Even if some of their questions may have indicated a poor understanding of the topic, these people were not discouraged since they had confidence in their intellect and abilities, and were not worried about how they were perceived. They asked them plainly, without trying to hide their lack of understanding, or trying to 'dress up' the questions by throwing in unnecessary acronyms or technical jargon.

Asking lots of questions, even if you have to force yourself (odds are that you are a natural introvert), is a wonderful way to build rapport and be 'on the same page' as the person you are asking—and it's a great way to connect with any audience you may have too. The expert will gain an appreciation of your thinking and they will respect your willingness to learn.

Learning to ask questions also takes some practice. With time you will give off a more genuine, inquisitive feeling that will result in people jumping fences to get you the right answer!

Pretend it is Your Business

One of the most useful approaches that I have found that you can use in the workplace is to pretend that it is your company that you are working for. This may not be necessary all of the time, and sometimes you may not have the flexibility in your role to give this a proper go. However, in most situations this attitude is very valuable—for example when you've reached a dead-end, or when you don't know how to proceed or how much effort to put into a particular task, which technical direction to take, what people to employ, what equipment to buy, what social events to hold and so on. When you pretend that these decisions are being made for your own company (and you will enjoy the spoils of good decisions, but also suffer the consequences of bad ones), decision making takes on a very different role. You suddenly acquire a sense of responsibility—a sense of urgency. You tend to be more active in your communication. Needless to say, your superiors will certainly appreciate your approach (presuming that you don't actually attempt to run the company, which might be overkill!).

I have also found this approach very useful in identifying what behaviours I should tolerate of the staff that I supervise—when I should be commending them on a job well done, and when I should be telling them off. In the past, I have found this difficult, but now I use my own internal standards as a reference. I pretend that I have hired them to work for my own company. If a staff member does something that breaks my own standards, then I know that they aren't performing well, but if they exceed these standards then I know they are performing well.

This mindset also prepares you for more senior roles, since you not only develop a sense of responsibility and due diligence, but you also begin to place effort into defining what your own standards are (and you are in touch with those all-important values of yours), and what direction the company should be taking. This will come through at meetings when your opinion is sought. It will be clear that you have taken the time to ponder the 'big picture' and most importantly, you genuinely have the best interests of the company in mind.

Exercise 7: What would you do differently if you were in charge of your company? Write a paragraph describing what your workplace would look like (what projects and products you would be working on, what people would be working on your team and what culture you would encourage). If you are still studying, think of your university or college degree.

CHAPTER 3

Achieving Outstanding Success

By this stage you should have an awareness and understanding of your values, as well as some goals that are aligned. You have a set of strong, consistent and reliable beliefs that won't fail you and will keep you on course. You understand that there will be endless hurdles, but they are working for you— they are opportunities. You now know not to be discouraged by these hurdles, like most people are, but expect to see them and know to plough through them. You know that you must try new approaches and change direction as necessary. You know the importance of rewarding and congratulating yourself. You also have a general understanding of how the engineering world will test you and reward you. Finally, you now have an idea of where you would take your company or university if you were in charge.

The question you may now be asking is how can you keep moving forward? While the first two chapters have laid the foundations for success, there are other considerations you need to be aware of. The first important enduring trait is maintaining motivation and momentum. It is natural to have periods of high motivation where progress seems easy, and periods of low motivation where progress seems impossible. Everyone has these peaks and troughs throughout their engineering careers. The important thing is to plough through the troughs. Don't get stuck in a rut—keep moving forward. This is why it is so important to have an emotionally compelling reason (that is, your 'big picture') to do what you are doing in the first place. Establish this immediately! Make it clear in your mind, and review it daily, at least until it becomes second nature.

Exercise 8: What motivates you to pursue your career? Write down this grand goal. If you say "to make a comfortable living" or to "pay off the mortgage" then go back to the start of this book and reread the first chapter! For example: Do you want to help people? Do you want to create something awesome? Do you want to be the finest specialist of X on the planet?

Get into the habit of making a plan for the month ahead, in terms of goals. What would you like to achieve by the end of the month? I normally further break this down into weeks, where I focus on more specific goals. I usually do this on a Sunday night. You want to be getting into work on Monday morning knowing exactly what you will be focusing on and why, and you want to be doing it with a high level of motivation. Then tick these accomplishments off as you progress through them. You should be getting out of bed most mornings with a high level of enthusiasm, knowing what's in store. It is best if you can get to a point where you are looking forward to every single day. What exciting challenges are you going to tackle and what will the outcomes be?

> Exercise 9: Make a game plan for the next thirty days. What would you like to achieve? Break each goal down into constituent micro-goals. Include learning goals. After the thirty days, assess how much you have achieved and repeat for the next month.

Throughout these thirty days, try to learn something new every day. That is, incorporate a learning goal. I'm not a fan of learning for learning's sake, but if you've identified an important field, and one of your goals is associated with learning, then incorporate this into your planning. For example, you may have learning goals such as learn programming language X, improve my writing skills, improve my public speaking skills and so on. It is best if they can be broken down into very specific constituent goals—otherwise you will never know if you are moving forward!

As I've stressed before, it is essential that you write down your goals. It is even better if you tell someone else of your goal since it adds to the pressure of achieving it. Most large organisations have a performance management and assessment scheme that you can utilise to lock down those goals. This can be effective since it commits you to what you want to achieve, as well as making it clear to management that you have a career plan and career awareness—something that is normally seen very favourably. However, beware of having your goals driven (i.e. limited) by others. These schemes can be limiting in terms of what you can add to them (and you may not have the flexibility to add what you would really like to add) so use these tools with caution. Your grand goals should never be at the mercy of the decision makers in your company, regardless of how competent they are.

I will reiterate that you should attempt to improve a little bit every day. This will give you a daily, micro confidence boost. For example, meeting new people is a worthwhile investment since you can learn from them and have someone to call on when you need help. Are there organisations related to yours that you could get involved with? Learning about your own organisation is another worthwhile investment. Perhaps unbeknownst to you, your company has a set of guiding principles, a vision, a mission, a structure, policies for everything from work, health and safety to procurement, and is subjected daily to external pressures and factors. Learning something new about your company should be considered as an achievement. Technology is yet another example. If you learn about a new technology or attend a seminar, this is another form of achievement and a form of self-improvement—something initiated from within that will quickly separate you from the pack.

Opportunities

Once you been working enthusiastically and effectively in a consistent direction for a while, doors to opportunities will inevitably open up to you. It is of critical importance to take advantage of opportunities. Realise that you worked hard for them, but beware—they are often disguised in very unexpected forms! They are still 'earned' and potentially extremely valuable. Firstly, it is important to be able to spot them (this takes some time and practice and is half the battle), and secondly, you must act in appropriate ways in order to fully take advantage of them.

Realise that opportunities are everywhere—you just have to spot them and take advantage of them. I am staggered by how many people tell me they have reached a dead-end while I can see numerous opportunities right in front of them. It is as if they have been blinded. They have either accepted that there are none, and this is what they see, or they have lost the desire to look for them. Perhaps they can see them, but they don't appear as opportunities. It is also amazing how many opportunities are squandered every day. In such cases, the person can see the opportunity, but he or she doesn't recognise its true value, or doesn't believe in themselves enough to pursue it, or perhaps they are discouraged by the potential work involved in taking advantage of it.

Exercise 10: Think about an opportunity that you allowed to pass you by recently. How would your life be different if you had capitalised on it? Think of an opportunity that you took advantage of recently. Think of how much better your life is because of it.

Two essential elements that you should focus on in order to begin making the most of the opportunities around you is that detailed plan from the second chapter, as well as those emotionally compelling reasons to want something badly enough. These are ingredients that will prime your mind for spotting opportunities. Armed with the other skills and tricks outlined in this book, you will have the ability to create, spot and capitalise on opportunities.

Don't be discouraged if this doesn't happen immediately. When it does happen for the first time, make sure to congratulate and reward yourself. You need to be given positive feedback to be on the lookout for further opportunities. With time and practice, you will develop the skills to create, spot and take advantage of opportunities. You will differentiate yourself from the vast majority of people around you since you will be on the hunt for opportunities—you will smell them a mile away and when you see them you will pounce!

I've heard a saying many times over proclaiming that everything you could possibly want is possible if you want it badly enough. This is something you should consider, since there is a lot of truth to it. Every person that I have met so far in engineering is operating well within their comfort zone. Furthermore, every person that I have met so far is not aware of their true abilities. People, particularly introverted engineers, tend to drastically underrate what they can achieve, or where their horizons can lead them. I think the problem is that true achievements are beyond your immediate horizon, and of course if you can't see something, it is hard to believe in. This is where you must have faith, which I talk about later in this chapter.

Confident, successful engineers who regard themselves highly understand that their true achievements will surely be realised, even though they can't see a direct path ahead. They plough on, working hard towards their goals, their senses acutely tuned to opportunities. They will not be discouraged, whether it takes two years, five years, ten years or more!

Continue to Persevere

I have talked about perseverance in the previous chapter, but this is such an important concept that I will reiterate it throughout this book. You only recognise the importance of perseverance later down the road. As a graduate, you feel that your path is limited. However, you must work like your life depends on it. This is another one of those lines that you should reread. If you attack your goals as if your life depended on them, I guarantee that doors will open and eventually you will reach your goals. *The engineering world is particularly rewarding for goal-setting people with clear vision.* As explained previously, engineering does an excellent job of hiding your rewards from you and tests you regularly to see if you are really up to it. This is why the vast majority of engineers give up trying to achieve their most important goals, even though they may be on the right path.

Don't be discouraged by the following statement: true accomplishments are much harder than they originally seem. However, they are achievable! This has been shown again and again by people who have started successful companies or made a massive impact in engineering and science. If you think it will take six months to achieve something, chances are it will take three years with many more hurdles than you think— but you will get there! Beware of this. However, every coin has two sides, and the positive side of this particular empirical observation is that you can generally achieve much more than you initially imagined. If you think that you can start a company with twenty people and $1M turnover, then chances are that you can start a company with a hundred people and $10M turnover.

In a nutshell, when it comes to major goals, chances are that getting what you want will be much harder than you think, but what you can achieve is much greater than what you think. Be aware of this and don't let this discourage you.

Hope Versus Faith

Throughout this book so far, I have talked about the importance of having faith. However, when I talk about faith I mean something very different to hope, which very commonly gets confused for faith. If you have put the work in; are now getting feedback that you are making progress (feedback is nice to have but doesn't always happen); continue to persevere knowing that you will eventually reach your goal; have positive feedback and are celebrating your wins; are confident that you are on the right path; have just changed your approach to avoid an obstacle; or if all the indicators are right and you are moving forward, then having faith that you will eventually reach your goal is healthy. In fact, it is essential. Sometimes your destination is obscured so much that it is only faith that is driving you. In fact, I would say that throughout the majority of your engineering career this is the case. You may, of course, be getting different levels of feedback. I will discuss what the different forms of feedback are later in this chapter.

Blind hope, on the other hand, is undesirable. Let's say that nothing is working out, you have no idea how to proceed, you have stalled and you are hoping for the best. You have hope that it will all turn out fine. It won't. In this case, hope can be problematic since it will ground you and prevent you from taking any meaningful action. In fact, this type of blind hope can waste years from your career since this also applies to career planning. If you are unhappy in the workplace, unhappy with the people and unhappy with the pay, then hoping that it will all change is insane. Situations generally do not change unless you take action to change them. If they do change by themselves, they can take a long time, like ten years, and by that time you have other priorities—career opportunities have passed you by, and it is unlikely that you will get exactly what you were after in the first place anyway. With this in mind, never just hope for the best!

Faith, which summons you to take action and guides you to your destination, is something very different and something that is essential—otherwise we would never make any significant progress! It is a consequence of having confidence in your skills and clarity in your vision. When you become skilful at a particular task, you just know when you are headed for success even though others around you may not.

It is important to remember that problems just do not work out for themselves. This is one of those lessons that took me a decade or more to work out. Over time you will learn when to have faith. Generally speaking, when you can see the work going in, and you can see yourself heading somewhere, and you are getting positive feedback (for example, you get a raise, or a project succeeds), then faith is healthy (as I've already mentioned, I talk about feedback mechanisms in detail later in this chapter).

As I've said before, sometimes indicators of success can be hidden from view. Paths towards success are almost always indirect. For example, you may be striving to win over client A, but in the process you get an award from sponsor B. They do not appear to be related. The important thing is that you have a goal; you are moving forward, and you are being rewarded. The exact path may be unknown until you get to your destination!

As a rule of thumb, the moment I feel hopeful, I also feel fearful. At this point, I know that something is wrong. I may not know what it is exactly, but chances are either myself or my project is headed for trouble. Hope never achieved anything in engineering. Hope is your enemy. Hope is the result of fear, and if you are fearful then it is your intuition telling you that you are about to fail. Never be hopeful of anything in engineering—it is almost as counterproductive as giving up.

Luck

If there's one philosophy out there which causes a lot of heartache and is extremely widespread, it is the notion of luck. We are so used to looking at people, companies and projects that are successful and attributing it at least in part to 'luck'. Of course, sometimes we know they worked hard for it but there is always an inner thought that attributes it to being at the right place at the right time. Likewise, when something undesirable happens, we are taught in society to attribute this to 'bad luck'.

The more positive and negative things that you experience in engineering, the more you realise that there is no such thing. Even at times when you are absolutely convinced luck is at play, it is not the right point of view to take.

Get used to thinking—even if you have trouble believing it—that you create your own good luck and your own bad luck. Even if in the very few cases where you truly couldn't have had any influence, it is unhealthy to believe in luck. Studies have shown again and again that successful people never regard luck as a factor. They may say this in passing, in order to brush off their successes. A sportsman may say after winning a game, "I guess I was really lucky!", but they know it is far from the truth. They are just being humble. In some cases, it may even seem like luck to them, but the reality is that luck had nothing (or very, very little) to do with it.

Conversely, unsuccessful engineers regularly attribute their failures to bad luck. They will say, "It wasn't my fault, I was just unlucky!" or "It was just bad luck that my project failed." When something works for them, the same people tend to attribute their successes to good luck, rather than accepting that they probably had a lot to do with the success! They will also attribute the successes and failures of others to luck. They will say, "He was so lucky with that customer! Who would have thought that she would be from the same town?" or "She was so unlucky to have missed out in that round of promotions!"

The difficult thing to accept is that success and failure are very cleverly disguised as good luck and bad luck. Unless you know this, you are likely to put all successes and failures down to good luck and bad luck respectively. The problem with this is that when you truly believe that events are out of your control, then you are less likely to take corrective action to prevent your failures and affirmative action to reinforce your successes. By not reinforcing the behaviour that got you the success, by not praising yourself following the success, you miss out on a valuable confidence recharge in the form of positive feedback.

Realise that you create all your good luck and bad luck, no matter how disconnected the outcomes appear from your actions. This means that when projects are not working out, then take corrective action, and persist. Of course, I am not saying that you should get down on yourself. Getting down on yourself excessively will only result in you getting demoralised, demotivated and may lead to you quitting the project or losing motivation. Accept what has happened, accept that it was probably your doing, learn from it as much as you can, realise that mistakes happen all the time and they are generally healthy, then move on. Likewise, when you are successful at something, realise that it was probably your doing! Praise yourself, smile, laugh, savour the moment and move on!

Playing by the Rules

No matter whether it is the workplace, in the sports field, or in everyday life, all of our relations with others and our behaviour is always bound by a set of rules. The trick to mastering your professional life (and for that matter, your sporting life and personal life), is achieving what you want to achieve without extending beyond what is legal, ethical and proper. In your journey, you don't want to be hurting or insulting people, or cheating, stealing, blackmailing, taking credit for other people's work and so on. I'm sure you are very familiar with the rules of our society and you know what I mean. You may also believe that this does not apply to engineering, but it certainly does! You will be tested and tempted again and again, just like on the sports field.

The truly successful people can achieve what they want without breaking the rules and without causing pain to others. It is a mistake to think that success has to come at the expense of others. You are generating value and serving people through your successes. Your rewards are recognition of the above. Is there anyone that you aspire to who has achieved something great, someone who has made their achievements because of a dishonest or unlawful approach? I suspect not. This is simply one of the challenges of life, but it should be seen in a positive light, like the lines on a football field. Accept that they are there and learn to live within the lines. The rules are there to keep society functioning. Realise that if they were not there, society would be back in the dark ages—incoherent, unguided, unfair, unrewarding and dysfunctional. To move society forward coherently and fairly, there has to be rules and boundaries. It is no different in the professional world of engineering.

Now, I'm not saying that you shouldn't push to the very limits of the rules, and question the rules, and look for caveats and sometimes even stretch the rules. I'm not saying that you shouldn't be aggressive and bold, pushing the boundaries in everything you do. At some stage in your professional career, in order to achieve what you want to achieve you will probably insult and upset people, argue with and ignore people, sack people, have people cry in front of you and swear at you. You may occasionally be warned by your boss for pushing too hard, just like occasionally you may receive a parking fine or two! This can be expected with any driven, successful person since it is part of the learning process and your attempt to locate the right boundaries.

At the same time, I'm not saying that you should be someone who is easily walked over—someone who tries to please everyone. I'm also not saying here that you should always take the conventional, well-proven road, and remain right at the centre of your comfort zone. Far from it! I discuss the importance of being original and driven later in this book. But don't act at the expense of others. If what you are doing is unfair, illegal or unethical in any way, cut it out and find another path— there is bound to be one.

Another way of appreciating the importance of rules is realising how it would feel if someone else broke them at your expense. For example, let's say you have settled down to a rhythm, found your form and you are doing what you love doing. You have a company and you are making a wonderful product and producing a wonderful service. You have done everything right. Someone comes along and robs you of all your effort—very unfairly and unethically and perhaps even illegally (this is a completely hypothetical scenario—this has never happened to me!). Everything that you worked for is gone. How would you feel? It is very different to suffering a loss brought about by your own mistakes. You feel truly gutted—like the world has caused you great harm. You would lose faith in the people around you, since you would feel 'ripped-off'. You would lose trust in society and the world, and you would be less likely to succeed and push the boundaries in the future. You would become defensive and constantly on your toes. You may even be tempted to demonstrate the same type of behaviour against others. It is fair to say that this sort of malicious act hinders society—its effect can be magnified significantly by the psychological effect it has on people.

So what exactly do I mean by the 'right thing'? Not all behaviour is defined by a set of rules. Where there are no written rules and regulations, develop your own ethics and your own standards of behaviour. Don't be afraid to change and adapt these as you develop as an engineer. You will naturally find that the rules that you play by change—perhaps they will become more restrictive causing you to become more self-centred or perhaps they open up and you become more philanthropic.

There are several reasons why you should always play by the rules. The first one is a clear conscience. It is very important to have a clear conscience and maintain good 'karma'. If you feel that you have been unfair to someone, then it will ride on your conscience for a long time, and you may not feel fully free to push the boundaries again. Realise that it is not only others that you are competing against. They are the minority. *As strange as it sounds and as difficult as it is to comprehend as a new starter, the majority of your competition is against yourself!* Realise that you have a mind that has biologically evolved to battle with itself. This must sound far-fetched or even ridiculous for a graduate of engineering. The bottom line is, if you are carrying around guilt wherever you go, you are less likely to be successful, since deep down inside you don't feel like you deserve to be rewarded. In fact, you subconsciously feel that you deserve to be punished! How many talented sports stars do you know who seem to always fail under pressure, at the all-important final stage? If you believe deep down that you deserve to be punished for something, then you will act out in a way that attracts that punishment, through failed contracts, poor sales, poor relations with people and so on. You want to be going into every situation and every project with the mindset that you *deserve* success—that you have done everything right to justify and attract success!

Secondly, if you succeed in being deceitful without getting caught out, then you are likely to continue with your behaviour since you have just been rewarded. Your mind senses these rewards and you learn to repeat the same kind of behaviour. Getting something for nothing is a terribly addictive way of operating. Once the rewards come cheaply, you begin to see the gains as an entitlement. It is perfectly fine to be driven (in fact, this is essential) but not to gain from what you have not earned. Unfortunately, there is no upper ceiling for bad behaviour. You are likely to magnify the habit, particularly when you encounter difficult times, and it will become worse and worse, until eventually it is likely to have major ramifications. Eventually you will get caught out. It is very difficult to get your reputation back with clients, employers and within an industry if you have seriously messed up. Generally speaking, it takes many years to build your reputation, but only one slip-up to ruin it.

The third reason why you should always play by the rules is that the type of behaviour that you subscribe to (for example, lying, cheating or stealing) tends to come back and bite you, but tenfold. The universe has a funny sense of humour, and your behaviour will tend to return back at you, with a twist. For example, if you have been selling a client short for years, often you find that the client ultimately sells you short in an unexpected way (again, this is just a hypothetical scenario—I have not done this!). Or perhaps your next client almost causes you to go out of business because of some sneaky contract clause. By demonstrating a behaviour, you are effectively saying to the world: *"I believe this behaviour is perfectly acceptable. I am comfortable with this behaviour and its consequences from both ends. In fact, I think that everyone should do it! I would be happy if others did the same to me."* Unless you can comfortably and honestly say this to yourself, then don't do it!

Finally, you will be surprised at how much of your deceitful nature will actually show through. The people who succeed in business are those who can be trusted. If you repeatedly lie, then you will develop a habit, and you will exhibit subtle signals in the form of facial expressions, body language and verbal language that will give away your deceitful position. We've all met people who we just knew we could not trust. We've learnt from a young age how to interpret these signals—it's in our best interests to do so. Most of the time we don't even think about it. We just get a particular feeling when someone tells us something. When we meet new people, we quickly and automatically attempt to decipher just how genuine the people are. Of course, sometimes you can't trust these initial gut feelings. Sometimes these people turn out to be trustworthy. Sometimes other people, who give off all the signs of being genuine and transparent, turn out to be quite the opposite. However, in most cases, your gut feeling is right. In any case, it is difficult for you to do business with or trust someone who is giving off untrustworthy signals. Therefore, if you lie or are deceitful, you are likely to give off undesirable signals. People will have difficulty warming to you, and they will be reluctant to accept your proposals.

Doing the right thing will foster trust in your colleagues, supervisors and clients. When you become trustworthy, your ideas will be taken more seriously and you will win respect. This is particularly important when you are young and haven't proven yourself to your employer yet. Of course, there are many ways of winning respect, but I regard a genuine, open and transparent nature to be one of the key characteristics that you should exhibit as an engineer. It is also difficult to execute many of the other principles that I discuss in this book without being honest and trustworthy. I'm assuming that you have earned your right to practise as an engineer (or are about to), and you have done so honestly and have nothing to hide.

Don't ever be drawn into the deceitful behaviour of others, whether it is real or just perceived. It is very easy to observe someone else's behaviour, and use it as an excuse to do the same. You can easily come unstuck, regardless of whether it is actual deceitful behaviour or just perceived. The ramifications can be identical. *You must be strong and resist the social pressure of being drawn into unacceptable behaviour.* You must not let yourself be drawn down to another person's level, whether it is a conscious decision or a passive one. Dishonesty inspired by someone else is no different to dishonesty from within—it has the same potential to grow out of control and damage your career. If you see it, then address it or get out of the way. The more time you spend around deceitful people and within a culture of deceitful behaviour, the more likely you are to succumb to this behaviour. Beware—eventually you will!

The final reason why it is so important to be honest and do the right thing as a habit again relates to the magnitude of your possible achievements and the extent of your horizons. There will always be other engineers who are higher up the ladder than you (perhaps they are smarter or more successful or perhaps just more senior) who are watching you. There is always someone assessing you, even if you are convinced that there isn't. There is always room to improve and to grow naturally, without breaking the rules. You will always be tested, and your actions will always have ramifications, no matter what stage in your career you are. Therefore, there will never be a time when it will be acceptable to break the rules. In short, do the right thing as a habit, and treat others and the rest of the world as you would like to be treated.

The Model Employee

It is ideal to have an idea of what to strive for in the workplace. What sort of employee do you wish to become? The best way of understanding the model employee is to put yourself in the shoes of your employer. They want someone who is a self-starter; someone who can obviously do the job well and will make them look good; someone who doesn't complain and who needs minimal supervision; someone who develops and grows with the company; and someone who always maintains high levels of motivation. They generally want someone who is a professional, doing what is required of them even if they don't feel like it or feel it is beyond their immediate capabilities.

At the same time, your employer will normally be more than happy to watch you grow—to invest in you and to nurture you. It is natural for a reasonable employer to want to see you grow, evolve and take joy out of working for them. You should never develop a 'me versus them' mentality at work. You should realise that you are in a mutual agreement, like a marriage perhaps, where both parties are there to grow and benefit from each other. All you can do is work on your part—what you have control over.

You should be creative and driven—willing to present your ideas and to pursue them. You should strive to be independent. Very few bosses like spending more than several minutes a day correcting and guiding their employees. They are normally too busy to devote more time than this.

You have to be persistent. *No boss likes having problems return to them.* In fact, you should think twice about passing problems back up. Often they are well inside your capabilities. There is, of course, a fine line between what you decide to persevere at and what you decide to pass back up. But there is a big difference in asking for help and returning the problem to your boss! Asking for advice implies that you are eager to continue with your task, but just need some assistance. Passing the problem back up implies that you believe the problem is too hard for you. Not only that, it signals that you don't want to deal with it, and think it should be your boss's responsibility. This is not a healthy attitude.

I've also commonly seen a 'I'm too good for this place' mentality. Even if it is true, you should use your abilities and knowledge to move the place forward. Make suggestions and offer solutions that require the least work on your employer's part. Don't expect an employer to jump at your ideas, but just because they don't jump at them doesn't mean they're bad. *Realise that everyone has drastically different viewpoints, and it can be very hard to sell your ideas even if they are potentially of immense value to the company.* In fact, the ideas that have been most valuable for me, leading to the most successful projects, have actually been those that were the hardest to sell and were repeatedly rejected by management! In other words, don't expect others to see gold as gold. Realise that your management are busy and worried about other problems. Often it is easier to remain with the status quo rather than committing to a new starter's ideas. Don't think that anyone is out to get you. Don't think that a boss simply doesn't want a new starter to outshine them. These are the excuses that poorly performing graduates use to account for their lack of progress.

Similarly, when you see others get their project ideas through, don't think that 'they were friends with the boss' or 'they had a fancy degree' or 'they are more popular'. Chances are they got their ideas through because they were better ideas and they were more effective at convincing management or clients of their value. Even if any of the above were true, it is not a healthy point of view to take. In the vast majority of cases it simply isn't true, and in those rare cases where there is an element of truth, then just learn to live with it. Don't allow the situation to diminish your drive and the faith in your efforts being eventually rewarded.

The worst thing you can do is resign or stop caring, believing in your superiority over the company, and believing that you can't change 'the system' because of inherent flaws in 'the system'. I've seen this attitude and it is not healthy. Accept also that sometimes the time is not right for your suggestions. Perhaps they need to mature. Perhaps the right opportunity has to come along. Also be aware that *most people are naturally opposed to change*. Some people are so strongly set in their opinions and so determined not to change that it is difficult to comprehend. This is despite numerous alternatives on the table which may be blatantly obvious to others.

The ideal employee is loyal. You shouldn't be criticising the company, or selling the company out, or jumping ship too quickly. Even thoughts along these lines show through in one's character as a lack of loyalty.

Be honest. *If an employee is ever dishonest, it causes an immediate loss of faith in the employee by their peers and management which can be hard to rectify.* It takes significant time to undo this type of error. Don't lie about what you did, or what you were planning, or what you knew. If this dishonesty is aimed at hurting your management, you can kiss a successful career at the company goodbye. And don't ever underestimate the power of word of mouth. The bad reputation that you establish at one company will quickly proliferate across the entire industry. It never ceases to amaze me how quickly gossip can spread across an industry.

If you lie outright—that is, tell your boss you did something which you didn't do—then you might as well wave a successful partnership goodbye. In almost all cases, it is better to own up and admit your error, and profess it won't happen again, than to lie. I have seen some shocking mistakes in the workplace (and I'm not necessarily talking about the engineering workplace here—everyone's had a terrible experience getting their car serviced, or with a real estate agent, or perhaps with a tradesman or professional). In most cases, if the person owns up to their mistake, and is sincerely apologetic, then their wrongdoing may even be overlooked. Anyone who has achieved anything great or is in a senior position understands that everyone makes mistakes, regularly. However, it is what you do after you have made the mistake that matters. If you lie, cover it up, or blame it on someone else, then in the long run you will suffer. Not only will you be eventually caught out, but you will develop an ingrained habit of avoiding blame. You will also not learn from your mistakes as effectively.

Finally, *never make your employer look bad*. This is such an obvious point, but I still see employees of all levels purposefully laying the blame on an employer when it really wasn't the employer's fault (and in cases where it was their fault, then diplomacy goes a long way). This causes management's level of trust in you to falter, drastically. It only takes a single slip-up to ruin your reputation at a company. With the rise of social media, your reputation is more important than you may realise, even at work. People talk. A lot. Do the very opposite by trying to make your employer look good. Every boss or employer appreciates this. I don't mean going overboard and attributing credit where no credit is due, or giving fake or insincere compliments. That is not what I am talking about. I mean that whenever you have a genuine opportunity to make your boss, your management or your employer look good, then do it, as if it were your own reputation on the line. This is such a simple piece of advice yet only a small fraction of engineers at any level do this.

Find a Role Model

It has been repeatedly shown that people who possess strong role models are more likely to be high achievers. It is important to have someone to look up to since it gives you a reference point for the type of attributes that you want to aspire to. It doesn't really matter who it is. It is best if you are genuinely driven and fascinated by the person, rather than just admire them because everyone else does. I'm talking about people who themselves have achieved something or exhibit a quality that you appreciate and aspire to. You can even aspire to people who have something that you want. It may not be clear to you yet how they got it, but with time you may be able replicate what they have done. *It is infinitely easier to make progress if you have a role model you can model your behaviour on.*

It is healthy (in fact, essential) to accept that there are people out there who are more successful than you. There are a few other reasons for embracing this realisation. Firstly, you become skilled at identifying which people can help you in various ways. Although I will cover the importance of people in your professional life in Chapter 5, at this stage in the book you should simply know that people are key to achieving what you want. There is a wealth of knowledge out there, far more than you could possibly comprehend, but it can be difficult to find and apply. Secondly, having role models makes you humble— you realise that you have a long way to go in achieving your engineering dreams. Other successful people will take you out of your comfort zone. You may think that you have arrived and you have nothing else to learn or improve on. A role model will encourage you to learn new skills. You will become hungry for new information and for learning new approaches, and adopting new techniques and viewpoints.

There is nothing worse than a new starter who thinks he or she knows everything, and doesn't aspire to being anyone. They are in a high-risk situation, since there is little motivation for them to go anywhere. Perhaps their brilliance, drive and other characteristics that they may possess will get them somewhere. However, this progress may not be in the direction that they want to go. People tend just to go with the flow until engineering becomes painful enough to warrant making a change. With a strong role model, this progress can be focused and the path taken can be more appropriate.

Look for role models wherever you go. What I mean is, hunt for behaviours that you aspire to or that you find admirable, whether you are in the workplace, getting your hair cut, or in the supermarket. There are exceptional people everywhere and if you can learn a single thing from a stranger every day, you would be an extremely capable person. Look at the high-flyers in your industry. Who is carrying out the creative or impactful work? Who is getting novel ideas out there? What are the companies that you most admire? What about outside your industry? Are there people who are really pushing through and achieving? Write down their names and their company names. Become more familiar with them. They each have a story and I can guarantee you that their journey wasn't easy! This process will eventually make it clear to you what sort of traits you value, and who can help you acquire them. Most people are fumbling in the dark. They have no role models and don't aspire to anyone other than the occasional sportsman.

Exercise 11: List five people that you aspire to and briefly write (one line) what it is about them that inspires you.

I find that I am constantly learning from people—not only more senior than me but also much more junior than me. Every now and then someone does something special that really impresses me. They don't have to be from your area, or even in your profession. For example, once a subordinate delivered a report way ahead of schedule simply because she had time (whereas the ordinary person would wait much closer to the delivery date, and hand in the work then). This really resonated with me and now I incorporate this into my own work plan whenever I can. Or when a car mechanic double-checked and triple-checked a bolt on my car, or when a technician thought about the big picture, and realised that what he was asked to work on would not actually be compatible with other systems. He flagged it as an important solution and raised it to management. Or when a researcher was proactive and completed a literature review which was sitting on my desk before I had even asked him! He knew what I was going to ask of him, and he anticipated and did the work, knowing that it wasn't a huge investment but would buy us some time and make the project flow more readily. Such events happen every day. Rather than getting envious when someone does something better than you had imagined, rejoice and steal (I mean, reuse!) their behaviour. When you reapply it later in the future, you will equally impress someone else passing on the behaviour. Most importantly, you'll be more efficient, and you'll gain the satisfaction knowing that you've done a good job and that you've improved!

Exercise 12: List three behaviours or three cases you have experienced in the last month that you aspire to. For example, perhaps you were being served a coffee and the staff were super-cheery and polite, and really made your day. Perhaps you were getting your car serviced, and the mechanic managed to locate a rogue part for you through his numerous contacts. Think of three cases where you were really impressed by the way someone did their job, and note what it was that impressed you.

Finally, if you are serious about leveraging off others—and you should be—then you should find a mentor. There are two great benefits of mentorship. The first is that a third party will keep you in line and prevent you from procrastinating and not attending to responsibilities. The second benefit, particularly if the person is from your field and more senior and experienced, is that you can benefit from their wealth of experience so that you don't make the same mistakes that they made. *Having a mentor can easily save you years of time* (I've heard that a mentor can advance you ten years in your career and I believe this). Though they don't have to be from your industry, there are some added benefits of having a mentor from your field since they will be able to offer more specific advice.

Generally speaking, seek advice from those who are much more successful and advanced than you are (who have done what you want to do), not those who are mediocre (who have not done what you want to do). But beware—often it is the people who have not achieved what you would like to achieve who are quickest to offer their advice. If you find someone who has done what you want to do, then it simply becomes a matter of duplicating their behaviour, which is almost always much easier than paving the way for the first time. Conversely, if you obtain advice from someone who repeatedly fails and has not achieved what you would like to achieve, then they will lead you down the path of failure without even realising it!

In order to find a mentor, you have a few options. Ideally, you should aim to find someone in your field who has already achieved what you want to achieve or exhibits those traits that you aspire to. Alternatively, someone simply more senior and more established is also a good option. These are the ideal options, since it potentially gives you all of the benefits of mentorship. In fact, the benefits that you can gain from someone who has already done what you would like to do are immense. Therefore, as I mentioned earlier, looking for and assessing people like this should be a part of your day-to-day approach (I discuss this further in the next chapter). If you genuinely aspire to being in this person's shoes and you genuinely admire them, then this can help a lot. People like talking about themselves and being made the focus of attention. Most people are also genuinely inclined to help you—particularly high achievers, since they are the most likely to appreciate your drive and determination, as well as your struggles. The majority of them will also understand the importance of career advice. If they are well known and very successful, it can be difficult getting their attention and time. In this case, I tend to look at them from the same level. Don't praise them too much and say that you are in awe. Just present the situation as it is. Simply state that you would like to get some career advice related to engineering. Often just a simple question can be a good first step.

The second option is to find someone from a different branch, company or profession who can mentor you. In this case, the focus will obviously be less placed on your particular profession, but more placed around achieving your milestones, dealing with people, staying positive and so on. You may even consider hiring a professional mentor. There are many available in all major cities. I find that the cost can be prohibitive for doing this on a regular basis (and you don't have as much freedom over the choice of mentor) so I normally pursue the casual mentor option.

Seeking out a casual mentor also expands your professional circles. Let's say that you talk to ten people with regards to mentorship (you don't even have to use the word 'mentor' since it may scare some people away—just say that you are after some career tips). These then become ten important people that you get to know, and they become intimately aware of your goals and will be on your side, potentially scouting for possible job openings and speaking of you favourably.

Make sure to be selective—don't just find a mentor for the sake of having a mentor. Selecting a mentor is a bit like selecting a spouse—the vast majority will either not be interested in you, or they won't be able to offer you significant value. Only a few percent of those you meet will be worthwhile candidates.

Also make sure to have a think about what you would like to get out of a mentor, or role model. Think about the types of traits and characteristics that you are looking to develop, and look for these in your mentor suitors. Have high standards. If you pick the wrong person (not that your commitment is necessarily lasting) then they can easily drag you down, drawing you away from your goals. They can make you more lazy, and encourage a negative mindset. You want to find people who are first and foremost positive. You want to find 'yes' people, who don't complain about how unfair the system is or how hard life is. Such a big part of your career is perspective and if the person only sees the negative perspective then this is where they will steer you! *Ask yourself, is this person bringing me closer to my goals or are they taking me further away from them?* In general, spending time around negative people who make excuses and complain is very infectious, whether you like it or not. I've heard before that you become the average of the five people you spend most time with, so take care when selecting your mentor.

Your mentor should possess traits that you aspire to, so that they pull you up. At the very least, they should have a neutral mindset but should acknowledge your goals, and have a desire in pushing you forward in achieving them. You should have a natural tendency to keep away from people who drag you down and discourage you. I will discuss this in detail in the next chapter.

The third option is to find a loved one or close friend. This may be the easiest starting point, particularly if you are shy in approaching strangers. Perhaps a brother, sister, father, mother, uncle, family friend or close friend would be effective. Perhaps you already know someone who is positive, driven, successful and who would be willing to help you.

Exercise 13: Identify three people who have the potential to help you as a mentor. Ask them for career advice and try to meet with each of them in person. Assess their suitability based on what I have said above.

When approaching a prospective mentor, be polite and don't be pushy. It is best if you treat the mentorship as informal. Asking for a formal mentor can scare people away, since they may feel it will be a long-term sacrifice on their time—something most successful people won't want to do, particularly those who don't know you well yet.

An informal arrangement works with you meeting, say, once a fortnight, or as necessary. Prepare for each meeting— have questions worked out. Have a list of aims that you wish to achieve as well as a list of some sticking points. This is particularly true for the first meeting, where you want to set a precedent and make it clear what you expect from the mentoring relationship. Preparation also indicates that you are serious and respectful of the person's time. As I've said before, you may not even want to mention the word 'mentor'. Simply ask an appropriately qualified person to meet with you to discuss your career over a coffee. It may help if you make it clear why you are asking that particular person. I normally mention some favourable trait of theirs and explain how it could assist an area that I wish to develop—people love compliments! For example, you could say, "I'd like to ask a few questions about leading small teams. I noticed that you managed to lead the last contract very successfully, so I'd be keen to get a few pointers from you."

I find that such an informal meeting usually works best. I normally don't mention mentorship with the person until I've sat down with them and had a very general career discussion. Be clear that you would like the discussion to last a certain time, such as thirty minutes. Make sure not to go over this. If you promise thirty minutes but don't stick to thirty minutes it will show that you are not true to your word—you don't value their time and it will establish a precedent for future meetings. They will know that you have a tendency to go overtime and may be reluctant to meet with you again. Remember, if they are a senior executive then their time is probably precious to them. At the end of your meeting, politely thank them for their time and advice. Say how beneficial it was to you. Stand up and shake their hand and say goodbye. You may want to thank them in an email a few days later too, but this is not essential— I only do this occasionally, when the person really goes out of their way, by lending me a book or referring me to a website or giving me something of great value that I can then follow up with. This is an ideal case since it sets up an ongoing mentoring relationship, where you start to bounce ideas back and forth.

You will find that a mentor will only be of value to you for a limited amount of time (usually one to three years). Believe it or not, after that time you will outgrow them, or at least grow in a different direction to them. You will no longer find significant value in what they have to say. This is perfectly normal! Then you should look around for a new mentor who can propel you to new heights. *Ideally, you will only draw out from each mentor the information you need at a particular stage of your career.* This is somewhat analogous to your friends and social circles outside of work. Sometimes, you just can't seem to meet each other's needs and grow apart—this is fine and natural, and nothing to get disheartened about.

Realise also that each mentor does have undesirable traits too, that you may not be aware of. They are people, after all! For example, if you have come to a point in your career where you feel you need drive and determination, then it is those mentors with drive and determination that will resonate with you most. You will be blind to their negative traits, and hopefully you will not take any on board.

Finally, you may also like to consider a group of like-minded friends or colleagues to create a mentorship circle. Finding such like-minded friends may be difficult, but you can be sure if you do they will support you and not drag you down. Meeting once per month may be sufficient. Perhaps three or four of you meet for a couple of hours, and take turns outlining your goals, where you are at and all of the barriers to success. Each month you meet for an update. The beauty of this method is that if you've selected them well, your friends won't tolerate you being lazy, procrastinating or betting on unrealistic events, as well as bringing in excessive emotion. If they are your close friends, then they may even call you out quite bluntly, which formal mentors may be reluctant to do. A good friend will always give his or her honest opinion, after all. In such an arrangement, you all benefit from each other.

Dead-End Jobs

The harsh reality is that there are many undesirable jobs out there. You could be bored out of your mind doing what you absolutely dread doing. You could be working with toxic people. You could be taken advantage of, doing a job that you really shouldn't be doing. Your health may be endangered. You could have a terrible boss, and I'm not even talking about anything illegal. More serious situations can leave you seriously scarred and fearful for your life, dreading having to ever return to work.

Accept that most employers don't have the intentions and standards in place that you would like them to have (I'm using the fact that you are reading this book to make a judgement about the high standards you are setting for your career). Very often, new starters will place people on a pedestal because they have some great skill, or experience, or a senior position. But they do not realise that these people are ... well, people. They have just as many flaws as anyone else.

Realise that people's negative attitude towards you is an indication of some inner conflict inside of them. You can't take things too personally. If you have an issue with someone, chances are someone else is having the same issue. Chances are that any reasonable person would be having the same issue. Don't be harsh on yourself.

The first important thing to realise is that it is probably not your fault if you are being mistreated. It is very easy, following a life in school, to believe that all of your superiors are perfect and you are to model them and bow down to them (well, perhaps you didn't regard your teachers quite that highly, but my point is that you were encouraged to respect them and look up to them, and in most cases this assertion was reasonable). If anything undesirable was to happen, then this was your fault. There is some truth to this, which I will explain, and generally a healthy attitude is one of accepting that your unpleasant situations are brought on by yourself. However, people are complicated and far from perfect. There are numerous situations that I have seen where the individual undergoing the ordeal did absolutely nothing wrong, but the situation appeared which made his working life a living nightmare.

How you react to mistreatment, bullying and harassment will make all the difference as to whether the situation gets worse, or you come out on top. There is a saying that we attract the way that others treat us. There is a lot of truth to this. If you are too malleable, or let people walk all over you, or don't stand up for yourself, then you will be trampled on and taken advantage of in some workplaces. If you allow for a toxic environment to pervade, then it will, and you would have indirectly played a part.

Most likely, you will land in a workplace where you will be generally happy. You will gather important experience, making a valuable contribution while getting along with most people.

You will notice the office politics and you may even notice some people having opposing interests, and even some tension and conflicts, but early on it generally won't bother you.

It is important to be open with your views, and make your standards known. Of course, this is hard when you are a new starter and you are only establishing your standards. This is why I believe those fundamental values that I described in the second chapter are so important. If someone, including a superior, breaks those rules or standards, then you should make it known and take the appropriate actions. For example, if you witness a person stealing company money (I'm assuming here that this would break your inner standards), then the person must be approached. If someone does something that you regard as inappropriate, for example makes an inappropriate comment towards you, then you should not let this slip. Breaking your standards will signal to them and others around you that you have no standards. This will attract more of the same behaviours until you rectify the situation. Furthermore, as more and more of this unwanted attention comes at you, you will find it increasingly difficult to correct, and you will have regrets for not addressing the problem early on. It is important that such problems are addressed as early as possible.

Tell the offending person that what they did is absolutely not acceptable. Have a chat to your supervisor. *Every inappropriate action or violation of your standards needs to have a twofold reaction and retribution—otherwise it will repeat again and again.* Now, I'm not talking about setting fire to the office! Be polite, calm and professional. Remember, if you don't take action, the situation will continue, and is likely to escalate. You don't have to be rude, or sound domineering or aggressive. The aim here isn't to offend anyone, even if you may have been offended. I am not saying that you have to face off with someone. Be diplomatic and polite. It is not about you versus them. Do not get drawn into an ego battle or a battle for superiority. Be the greater person, no matter what. Nothing will shake you, no matter what. Simply said, you have boundaries that are important to you, and someone has broken those boundaries. They need to be made aware and need to receive feedback telling them that this is not acceptable to you. In the vast majority of cases, people have no idea that they have offended you—they will regard their behaviour as normal. If it continues, they need to understand that you will escalate your reaction, either by reporting it to your superiors, their supervisor, or making a formal complaint to the company and beyond.

Keep a record of everything. Write down what happened, and when. Take advice. Talk to your friends and family. Keep everything in the open. Do not lock yourself down with the problem. Be proactive. You should overreact, but not necessarily in terms of a response in their face. Be rational. Get advice, and record it. Talk about it as much as possible. Do not bury it. Do not hope for the best. From my experience, the moment you 'hope for the best', you ensure that whatever it is you hoped wouldn't happen, happens again! If you don't prevent it from happening again, it will happen again.

There may be consequences of doing the 'right' thing, but unless it is going to cost you your life or your future, then my suggestion is to take action. *I have never yet regretted taking action, but I have many times regretted taking no action.* Displaying your high levels of standards will also show that you are a person of integrity and morals. It is also highly likely that others will respect you more, rather than less, from your managers, to your peers, to subordinates.

Having high standards will reflect on you in other ways that are indirect. By purposefully acting against major breaches to your standards, you will automatically react to subtle breaches, and hold yourself to greater account. As I said earlier, the offensive parties in most cases have no idea of the offensive nature of their actions. Assuming that you react appropriately, you are likely to ultimately garner their respect too.

Respect Your Colleagues

This should be an obvious one to all young professionals. Always treat your supervisors and superiors with respect. For that matter, you should treat everyone else with respect too, like your colleagues, subordinates, clients and associates.

This is not to say that your supervisors will always inspire or deserve respect. You may have some good ones, and you may have others that seem to be incompetent, detached, and even spiteful and condescending. However, most people are reacting to the emotion raging inside them. You don't know what is happening inside them and you are almost certainly not responsible. It is easy to think that you are at the centre of the universe, and every positive or negative outburst from your boss is a result of your actions. However, this is rarely the case, even if the attack is aimed at you for something that you apparently did or didn't do. Give them the benefit of the doubt. Realise that there are a million factors at play that you may not be aware of. Treating your superiors with disrespect, even if it is subtle, will result in massive black marks against your name in the long run.

Realise that your bosses are there for a reason. It is easy to think that they rose to the top through incompetence and cheating and leveraging from others or whatever else, but this is very rarely the truth and even if it was, it definitely does not help you to treat them as if it were. It is very easy to subscribe to office politics and gossip, and believe negative rumours about your management that probably aren't true. These rumours are usually spread by people less successful, who are stuck in their own rut because of their own decisions. You should avoid such people in any case, since they will only drag you down. If you absolutely can't genuinely respect your boss, then fake it! *Respect in the workplace is not something that has to be earned, even though this is in our nature. Respect in the workplace must be there by default.*

There is nothing more frustrating to me than meeting a recent graduate who is arrogant and tries to demonstrate to me how little I know. Furthermore, when I see a graduate who seems distrustful and lacks respect, it shines through. It causes an immediate loss of faith, since you feel like they will try to undermine you, and sell you out and speak negatively about you behind your back. If a subordinate treats you and your colleagues with disrespect by default, then usually they are not worth having on the team.

I personally apply exceptionally high standards of treatment not only to my superiors, but anyone older too, even if they occupy a more junior position in my company. This is a strange concept in some parts of the world but in some countries (parts of Europe and the USA) you always refer to older people as "sir" and "madam" or "ma'am" and treat them with higher levels of respect, as a matter of course. This does not come at the cost of treating people younger than you disrespectfully of course—you should still treat them with respect—but you may like to treat older people with even higher levels of respect. For example, when knocking on the door of a more senior person, I will always say "excuse me" and "thank you" at the end of the conversation. Whereas with someone approximately your own age it is fine to say "hey, what's happening?" or whatever you would normally say to a close colleague. Of course, you will have friends who are older who you treat like close friends, so this isn't a strict rule.

I have found that this philosophy works quite well in establishing a respectful workplace. It recognises the hard work and contributions that someone else has made, even if they have a frugal salary and junior position in your company. One day when you are in the same position (not necessarily in a low-paid position, but you will be old one day and there will be people who are younger than you who may be in more senior roles) it will mean a lot to you if they treat you with the level of respect that you deserve.

As I've already reiterated, you should also treat everyone else in the workplace with respect too. I don't mean that you should grovel or be unnatural. Form a respectful attitude as part of your natural behaviour. Sometimes you will be mistreated, and it is important to stand firm and not lose your cool. You don't want to explode and be dragged down. It is important to note that treating people with respect doesn't necessarily mean agreeing with them, or even doing what they want you to do. It means treating them as if they were as important or more important than you are. Don't underestimate how easily this shines through in your behaviour. I can immediately spot a loss of respect, perhaps caused by a change in approach or a decision that I made at a recent meeting. You give away your level of respect for someone very easily, through subtle cues in body language, choice of words and behaviour. If working with a genuinely high level of respect for everyone is in your nature, none of these subtle cues will be given away and you won't have to worry.

Never, ever make fun of anyone. This includes directly, or through other people. Even if you are joking. Some of it will always stick. There is a time and place for banter, but in my opinion it has no place in engineering. Be safe and steer clear of it. There are particular groups where banter is an everyday norm. Be aware of this since you may be faced with it. I can immediately spot a person who is going nowhere in engineering by the way they treat their colleagues. *No-one wants to promote a person who makes their colleagues or clients feel uncomfortable, even if they are highly competent.*

Never talk about people in a negative light behind their backs. It is one thing giving an honest, negative appraisal of a subordinate, where you are on their side and you are offering a solution; you don't wish them any harm and you want the best for them (in this case, think of them as a son or daughter if at all possible). The only exception may be providing a referee report on an employee whose performance has been disappointing.

But in this case, you should still be diplomatic and not let emotions or emotional language get the better of you. Give an honest appraisal, highlighting the positive and negative aspects, but in a positive light if at all possible. Steer clear of spreading gossip and office politics as much as possible. There is obviously good politics and bad politics. I am talking about the bad kind. Some level of 'good politics' is required in any business and can be healthy. Eventually, spreading gossip and criticising others behind their backs for the purposes of office politics will eventually backfire. This kind of behaviour has no place in engineering.

Don't take advantage of people and don't lie in terms of a project's urgency or in terms of your deadlines. You will eventually get caught out on this behaviour as well. Plus, if you treat all of your requests as very urgent, then people are unlikely to respond to you when something genuinely pressing comes along!

The Power of Saying "Yes"

Realise that some of the greatest things that will happen to you in your career (and perhaps in your life, for that matter) are the result of saying "yes" to something that you really didn't see coming—something you really didn't want to do, or you didn't aspire to doing. Your first reaction may have been to say "no", to save yourself the hassle and the headache. But then you realised that this could be a disguised opportunity (see the previous section on opportunities if you have forgotten what I mean). But really, it wasn't this thought process that triggered your enthusiasm—it was a deep-down value that wanted to help other people perhaps, a value that wanted to contribute, a value that wanted a challenge. There have been countless cases when I said "yes" to something, though I remember at the time being unenthusiastic and uncommitted, only for the project to ultimately flourish. The project changed hands, and another opportunity reared its head, and suddenly a massive door was blown open.

The occasions that I'm talking about normally happen when you feel indecisive about whether the task is worth doing or not. Obviously, if you are out of town or if you know you couldn't do the task in question, then this is another story. Similarly, there are requests that are blatantly unreasonable. I am not talking about jumping over mountains here. However, some requests that are merely challenging are different. They may be initially discouraging and often involve an unknown element or new people or a new area. Sometimes they happen to appeal to some inner value of yours. If this is the case, then this may be the smell of opportunity!

I am talking about those situations when you know you could give it a go, but inside you also know that it would be additional work, and you don't see any obvious, logical benefits (as discussed in the 'Receiving Your Reward' section of Chapter 2, some of the best rewards are shrouded from your vision—beyond your horizon). Often when approached you feel as if you are too important for the task (your ego rears its head, but more about that in the next chapter). You say to yourself "What does it matter if I don't do this?" and "It won't make the slightest difference!" and "No-one will even see this or care" as well as "I don't like Robert anyway! He's never done anything for me!" This last one is a particularly common trap of thinking to fall into. Don't get dragged down by others' reluctance to help you. You are not other people! You are different! At least I hope you will be once you have finished this book! Just because someone didn't help you out in the past, is not a reason why you shouldn't help them now. Perhaps you will start a new mutually beneficial relationship. Take note that you will naturally tend to search for excuses to give to the other person, such as "I'm not qualified" and "I don't have time" as well as "My boss won't be happy" or perhaps "I'm going away soon". These are generally not valid excuses. You know deep inside that you can push through. They are our initial, ingrained responses that we are used to giving.

Put yourself in the shoes of the requester. They probably have a real need to get something done. They have identified you as a person of high value—someone who can get the task done. They may not feel comfortable about asking you (everyone would like to do everything themselves, but they just don't have the time). They are not evil. They do not want to cause you harm. Perhaps they are in a real bind. Their project and reputation may be really hurt if you don't help them. At the very least, it will cause them dire problems. Most importantly, because it involves another person, it opens a new avenue for you (I discuss in Chapter 5 how important personal contacts are, but in a nutshell—you can slave away in your ordinary job for many years, but unless doors open for you and you learn to jump through them, you may not get anywhere in your career despite your hard work).

Furthermore, saying "yes" to others gives you the opportunity to prove yourself. Others are likely to see the quality of work that you perform, and may also come to you for advice. Realise also that chances are that the requester has already asked someone else, and undoubtedly the other person probably acted according to his initial reaction which was to say "no". It never ceases to amaze me just how similar people are—if your first response feels like a "no", then chances are it felt like a "no" to just about everyone else. It is times like this when you can differentiate yourself from the pack. This is the time that matters most—not (as I've said before) your grades at school or the quality of your suit or even your outright technical ability. This may represent a significant, career-changing opportunity that is awaiting you.

I can immediately spot a high achiever in the way they react to simple requests. Note also that how well you know the individual making the request should make little difference to your decision to help them. I've received requests from bosses who I knew very well, to people from different organisations who I didn't know at all, yet I've always tried to help out as best as I could. Obviously asking a lot from someone will often lead to a "no" answer and this can be expected. For example, if you asked your colleague to prepare a hundred-page report for you overnight, or clean your office, very few would say "yes" and those who did would probably raise some concerns in your mind! However, when the level of effort is reasonable, and it is targeted at the person with the right skills, then even though the goal may be dubious and the outcome uncertain (this is the worst case—often the goal and outcome is transparent), the high achievers will invariably say "yes" or find another way of meeting your request. It is as if they can see the potential of something happening—they can smell something opening up. They are excited by the unknown—by the challenge. I can immediately write off a colleague as lazy, pessimistic and frustrated if they repeatedly say "no" to my simple requests, which I know they could tackle with little effort. This response suggests they are inward looking, and they don't want to help others. This is not the right mentality to possess when you are part of a team. If I get the response "I'm too busy" every time, then I know it is a sign that they have no interest in helping me, or other people, and are likely to be self-centred individuals.

Be particularly mindful of what is seen to be 'cool' and 'uncool'. Trends are notoriously unreliable indicators of what to turn your focus to. In fact, one could argue that trends can be an indicator of what not to focus on. Take the share market, for example. When a stock is rocketing in a bull market, everyone is investing their money, and the levels of confidence are soaring. Unless you got in early, it is probably not a good idea to invest at this time. The same kind of trends are occurring in science and engineering. Technologies are coming in and out of fashion all the time. Sometimes these trends really do not have a solid foundation. Don't discount ideas that are not in fashion just because other people do. Make your own judgements. Do not listen to the masses or you will be often led astray!

If you happen to be the one making the request, it is important not to argue, get emotional or to plead. You can't change a person's mentality overnight, and not in such a drastic way.

When making your request, speak as if you expect to receive a positive response. If you end up receiving the assistance that you requested, make sure you give the person the appropriate credit afterwards. Thank them and tell them how much you appreciate their help, and how much it has benefited you. Speak favourably to your colleagues about it, and send a thank-you email to their boss. Make sure they benefit from the transaction as much, or more, than you did.

When you are the requester, you want to make sure that doors open up for the person helping you, and that in the worst case, their work is appreciated. I have seen this process work in both ways. Not only have I benefited from helping others through their requests, but I've also helped others greatly, by creating opportunities for them. Sometimes this can even happen inadvertently. For example, recently I asked a colleague to help out on a lengthy report. She attended meetings and did a fantastic job, gaining a reputation throughout the company. She was later offered a job in an adjoining section largely because of the motivation she showed on this task which was not her primary responsibility—it came out of an external request for help!

Indicators of Success

It is ideal if you are informed when you are making progress. Unfortunately, this feedback will be obscured and will come at you in various forms, as discussed in Chapter 2. When you receive this feedback, you must have faith that you are on track, and give yourself the positive reward that your mind and body needs. It is also critical that you receive a reward so that your motivation is maintained—or better still, increased—and you push on. Without these incremental rewards, you may feel like you are working continuously with no goal in sight. You may then lose sight of the big picture and not even remember what goals you are working towards.

Sometimes there are blatant indicators of success which demonstrate to you that you are on the right track. The key one that I wish to mention up front is becoming the 'go to' person. Generally speaking, it is good to be desired! Yet many young engineers misinterpret this, and don't draw anything positive out of this. They simply think that they are being taken advantage of.

In fact, there are a whole host of indicators of success that most people are not even aware of. You need to be, since if you don't know that you are on the right path, how are you supposed to stick to it? How can you improve without feedback? Negative feedback (through failure) is important, but positive feedback (from success) is also important since it reinforces the behaviours that got you the reward. You normally don't even realise how strong this reinforcement effect is, since it often happens subconsciously. When you learn to recognise the appropriate measures of success, then it will become automatic and your mind will just know that it did something right. Subconsciously, you will be primed to exhibit similar behaviours, which should bring you further success! If you want to make progress rapidly, then positive feedback should be used in conjunction with negative feedback.

So how do you know when you have been successful?

Sometimes this is obvious, but at other times it is far from obvious. Generally speaking, you know you are being successful when:

- You achieve an important milestone: you finish a task, you complete a contract, you finish a piece of code. This is an obvious one.
- Recognition and rewards: when you are being recognised by your peers and superiors through comments and official rewards and bonuses, then you know you are doing something right!
- You are being asked to do more things.
- You are being given tasks with higher levels of responsibility.
- You are being invited to more meetings.
- You are being included in more correspondence (for example, carbon copied to emails).
- You are developing a reputation for something. It may not be for something that you regard as a key area of expertise. For example, you could regard yourself as an excellent programmer, yet more and more people are complimenting you on your ability to attract customers. You think to yourself "this is not what I'm actually good at!" Don't worry if you are being recognised for something unexpected. This actually is of little importance.
- More people are hearing about you. You are being contacted by people you have never met before who would like your opinion on something.

When we were at school, our teachers always told us to give ourselves a 'pat on the back' for a job well done. This was actually the right philosophy to have. I've mentioned the importance of celebrating in the previous chapter, but it is worth reiterating here. If any of the above happens to you, then take the opportunity to give yourself a reward, ideally as soon as possible. If you finished a part of your project, then have a break and go and grab a treat. If you won a big contract, then treat yourself to a night out, or a movie, or do your favourite thing when you get home. Perhaps it is a video game or hobby. Or perhaps you like to cook a meal for your partner, or grab a beer with some friends, or eat your favourite kind of food. Do something that is rare and gives you a warm and fuzzy feeling. Ideally, tie an emotion in with the reward. Think of your success at work and where these successes are likely to take you. Associate all the good feelings that you are feeling with that success.

Limiting Beliefs

Once you are aware of how profound and widespread limiting beliefs are, your professional engineering life (as well as your personal life) will change forever. Once you are tuned in to spotting limiting beliefs, you will recognise them wherever you go—within your colleagues, within your managers and within your clients. You will hear them and see them everywhere, every day. You will also quickly realise that the people who tend to apply limiting beliefs are also the less successful ones, while the people without limiting beliefs tend to be more successful.

A limiting belief in this context is something you truly

believe is hindering progress, even though it does not necessarily exist. Although you can usually find some evidence to
support your claim (by selectively choosing and interpreting
evidence), a third party is likely to find flaws in your reasoning. It is therefore a belief, but as far as you are concerned it
is a fact. The important thing to realise is that it is so strongly
ingrained that to you it appears as a black-and-white fact so it is
very difficult for you to see the world in any other way.

Beliefs in general exist for a very good reason. By establishing your view of the world through a set of beliefs, the resources of your brain are not wasted revisiting ideas that have
already been proven to you beyond a reasonable doubt to be
known. These beliefs are built up over many years as you learn
about the world. In the majority of cases they are correct and
serve you well. For example, you know that your car will not
run on orange juice. You would not waste your time even con-
sidering this as a possibility, and you certainly wouldn't try it
out!

However, in some cases, beliefs can be extremely detrimental. For example, you may say "I will never get a job, since
I studied at university X, which is poorly regarded!", which
is a limiting belief. The big limiting beliefs grow over time.
Like a snowball, they gradually expand until they become so
well established that they are difficult to break down. Limiting
beliefs serve the purpose of giving you a rational explanation
for why you have not been successful at something. Worse still,
they give you an escape mechanism for avoiding the work you
need to do in order to achieve success. Once you have a limiting belief, you have a 'way out' of something. Why bother even

trying if you have this unshakeable wall in front of you? The painful realisation is that in many cases, limiting beliefs are not only self-inflicted (though sometimes they can be inflicted by others or society in films and movies, for example, which I will discuss later), but they are also not real! There may be an element of difficulty, but it can almost always be circumvented. A limiting belief is therefore an excuse not to take action.

Now, I'm not saying that anything is possible. You probably won't fly by flapping your arms furiously. However, there is always a way around a problem. There are much more effective ways of flying than flapping your arms. This is particularly true in engineering. There are always alternative approaches, many of which are far better than what you originally envisaged. You just have to find them. What makes limiting beliefs so problematic is that they force you to stop trying to find a solution! They represent the ultimate failure—from within!

Limiting beliefs are self-constructed brick walls that hinder progress. They stop you from going any further. Other people can certainly play a part in implementing and reinforcing your limiting beliefs, and for this reason it is important to always associate with the right kinds of people, particularly your closest circles of friends and acquaintances. If your friends tell you that something can't be done or isn't worth doing, then eventually you will believe them. Oddly enough, it is the people closest to you, whose beliefs you are most likely to trust, that can establish the strongest limiting beliefs. Spouses, family members and very close friends have good intentions, but they will fire off strong, emotional limiting beliefs that make you doubt your purpose. You have been warned! Furthermore, if all of your closest associates like to make excuses and carry limiting beliefs, then chances are they will infect you. Be on your guard and don't underestimate how powerful limiting beliefs can be! Luckily, once you become tuned in to what a limiting belief is, a warning siren will sound inside you whenever you hear one, so you will know to question it. Beware of the words "realist" or "being realistic" or "never" or "can't" or "impossible"— these phrases should automatically trigger warning bells. No-one ever achieved anything big by being realistic—ever!

It is only once you have had a limiting belief removed will you fully appreciate the damage they cause. You will also realise the power of tearing them down. You will wonder about other limiting beliefs and begin to question them too. Most engineers only realise a small fraction of their true potential. A major reason for not achieving more is that we fail to 'think outside the box' and one of the reasons for this is our limiting beliefs.

I have heard numerous limiting beliefs over the years. One employee felt he would never be promoted because he never attended the right university. "I will never be promoted!" he proclaimed. "No manager here has ever attended my university!" Ten years later, I met up with him again and he was still churning away, miserable in the same job. The reason why he wasn't promoted to manager wasn't because of the university he attended. The reason he wasn't promoted was because he wasn't creative and didn't manage his staff properly, and broke several critical deadlines. The reason why he performed so miserably was because his limiting belief told him he would not be promoted to manager, so he simply acted out accordingly. Had he believed that being a manager was a possibility, his future would have been brighter. Even if he hadn't been promoted to manager (in my opinion he would have been), his good work would have been recognised and he would have moved forward in some other way.

I have heard all sorts of other limiting beliefs, from "I'm too old" to "only evil people succeed" to race, background and education-related limiting beliefs. Here are some examples:

- I'm not adequately qualified.
- I don't have the right background.
- I'm too old.
- I'm too young.
- I'm the wrong race.
- This job isn't for me.
- Successful people are all <insert derogatory adjective>.
- Only <insert derogatory adjective> people are promoted.

Exercise 14: Have a think about what is stopping you from making progress. List three barriers that you think are stopping you from progressing. Think deeply about each one. Does it actually exist? Are you absolutely sure? What would it feel like if it wasn't there? Can it be circumvented in any way?

There is No Free Lunch

Accept that everything great has to be worked towards. Everything great that has been achieved is because one or more people committed heavily to its undertaking. Sometimes this is a direct reward which is blatantly obvious, but at other times the engineering world delivers the reward in unexpected ways. Perhaps years down the track—in a different guise or from a different angle—success finally comes. Sometimes it can be hard to spot how the effort and reward are correlated. I think it is for this reason that people lose sense of this connection. They attempt to 'cheat' the effort-reward cycle of the engineering world. In fact, I would say that the majority of engineers attempt to find a way of doing nothing in order to achieve something. Their whole mantra becomes one of doing the minimal work while maximising reward (the other part of this flawed mentality is the ego which we all must fuel, unfortunately. This is discussed in Chapter 4).

There is a difference, of course, in simply being efficient, and not redoing what has already been done and is available, to having a mentality or expectation that things can be 'won', like plush toys at a fair. It is a fine line and one that becomes clearer with time. It is important to:

- Not expect to get anything for free. It should be a nice surprise if you do.
- Be aware of your ego, and how much it demands to be fed.

Of course, cheating the 'system' may buy you something in the very short term, but you don't actually add any value to the 'system'. It may work once or twice, but it won't continue to work. You are using someone else's efforts and creativity. You are doing a 'trade' though you are really just borrowing. There is an argument that works conversely. That is, anything obtained for free has to be given back. There is a natural tendency for this in the workplace. If you keep on asking people to do things that are in your capacity of carrying out, eventually they will stop helping you. They will realise that you are a drain on their resources. Eventually, this may manifest itself in some of your own efforts to be drawn away by others. Realise that there is 'no free lunch'. Nothing can be gained for free, or at least the attitude of gaining results for nothing is not a healthy one to maintain.

This is something that everyone knows deep inside, yet very often people still expect to receive something for nothing. They see this as the ideal way of running their professional lives.

Realise that no-one has it easy, even if it seems as though they do. Everyone is tested over and over again. Everyone is faced with challenges daily, and they must find solutions. It is because you don't see how people react to these challenges (it is not documented in any way—you can't see it and you don't know what their action was) that you may mistakenly believe that other people have an easy ride. For this reason, it is possible to fall into the trap of thinking that only you have to work hard at something because of your misfortunes. This is because only your challenges are completely transparent to you. You can see what is required and the difficult path ahead.

Furthermore, everyone has their inner demons (I will talk about this later in the book). Everyone is tested with inner doubts. What matters is how they react to them! I will discuss positive attitudes and maintaining them in the subsequent chapter.

Stamp Out Problems

As an engineer, you are a problem solver—not a problem spotter or a problem conveyor! You are not adding much value to your workplace if you can spot problems but you are unable to address them. Don't just report problems to your supervisor, unless you are genuinely lost in terms of how to solve them. Similarly, don't allow your problems to bounce back to your supervisor or go out to your colleagues every time. The reason that you were given the assignment, task or problem to solve in the first place is that your boss didn't have the time to solve it. Obviously there will be times when you will need to report back to your boss and ask them to clarify the problem or ask them for advice. This is perfectly reasonable. I'm talking about something very different here which I used to do regularly as a new starter. Sometimes I'd receive a problem and realise it was tough. Rather than ploughing on, attempting to solve it, I would throw in the towel and report back to my supervisor, indicating that it was a show-stopper. For example, I would say:

- "That part is no longer available" or
- "He doesn't want to do business with us" or
- "He doesn't answer his phone so I can't get through to him" or
- "I don't have any experience with that".

Reaching a dead-end with a problem is not solving the problem, regardless of whether you report it or not. It is the very opposite. An effective engineer will find ways around an apparent dead-end.

Realise that the frustration you feel is magnified tenfold for whomever you bounce the problem back to. Your supervisor is busy with other tasks and does not want to be solving your problems for you. Realise that it is just as difficult for others to solve it as it is for you. Chances are that they would have hit the same barriers that you are hitting. It may seem like it will be easier for others, but this is because others have probably learnt to grind it out and persevere with their problems.

This applies also to returning problems in an indirect way. Believe it or not, there have been cases where I have given a subordinate something to do, and they have passed it on to someone else, and that person has asked me to do it! The task took a full circle and worked its way back to me! If you have been given a task, take responsibility for it and stamp it out like a small grass fire. Don't let small tasks escalate out of control. The only reason you should be going to others is when you are asking a subject matter expert for help—this is perfectly fine and can save you significant time. It will also hamper your reputation when your management sees that you aren't capable of carrying out your tasks to the end. Your supervisor, when handing you a problem, never wants to see it again. They want it carried out quickly, professionally and with minimal further input from them!

Be Original

I am often shocked at how little originality I see from new starters in engineering. There are exceptions of course. Every now and then a graduate is genuinely original, and is not afraid to suggest new ideas, but this is rare. I think one major reason

for this is the university system and the system of upbringing. Kids are taught to conform—to listen and follow. This is
a reasonable expectation as when kids are young they must be
disciplined to learn what they are taught, rather than go off on
different tangents, which are likely to be fruitless. Furthermore,
kids are normally taught in large groups with a group mentality. This automatically encourages a 'pack mentality' approach
to problem solving. This may be fine for some time, but there
must be a point in an individual's life when they separate from
the intellectual safety net of the herd, and of their teachers and
mentors. They must eventually step out on their own. This is an
unavoidable transition that must occur in every engineer's life.
In some it occurs while in university. In most it occurs after
they start working, and in others it occurs decades into their
professional lives.

In some cases, sadly, it never seems to come at all. I feel that
some engineers never experience this freedom, or 'engineering
liberation'. They follow a pack mentality until retirement. They
use the same formulas and the same rules. They don't realise
that in order to differentiate themselves from the pack and really
achieve anything spectacular, they have to think differently to
the pack. And not only do they have to think differently, but they
must be inspired to act and implement their ideas. These people
are unlikely to achieve true satisfaction in their professional
lives until they focus on those ideas that are inside of them.

This leads me on to the main reason for failure to express
and develop novel ideas—a deep-down lack of confidence.
Everyone at some point has a great idea, but very few will have
the confidence to act outside the normal set of behaviours. The

reason seems to be the pressure of conforming to social and industry standards. I see this attitude in research settings in particular. There is nothing wrong with pursuing someone else's area of research, but only if that is what really captivates you. But beware—it is too easy to fall into the trap of following the pack. It is also easy to justify such a decision with claims like "everyone knows that technology is fashionable" or "everyone else is doing it" or "there's a real future in that technology". However, sometimes this is just an excuse to take the easy road and not to make progress with your unique idea.

You should never feel inferior for going against the grain. I encounter so many young engineers who are afraid to publish something new because they feel embarrassed. This is why it's important to have your goals and plans developed. If what you are doing is consistent with your plan, then you will not worry what others think! Don't worry about being laughed at or 'banished' from your industry (within limits, of course; I am not saying you should lose your job over an idea). As long as your idea is authentic and consistent with your deep-down view of the world, then any negative consequences of getting your idea out there (there probably won't be any) really should not matter to you. This level of self-belief and confidence in one's ideas may take some time to build up, but you can start in small steps. I see the majority of engineers following the same approaches in their work and research, when often it is far from optimal. In fact, you would be shocked to learn how poor many engineering solutions really are, and how much potential there is for improvement. Graduates possess some extremely creative ideas—they just don't utilise them nearly enough, in my opinion!

When faced with a problem, try to think outside of the box. Try to think of new engineering solutions. Of course, there will be tried and tested engineering principles which will generally need to be applied first. However, there is always room for something new and revolutionary, even in the most mundane of tasks. It is far better to pursue your own idea and suffer the mild embarrassment of being told down the track that the concept already exists, or an existing solution is better. If you never try anything different then you are unlikely to improve on anything. In doing so, you are flexing your creative muscle and increasing your confidence in utilising creative ideas.

Be bold and courageous with your solutions and see them through to the end. It is one thing to have a novel idea, but it is another to make it happen. The old adage that 'achievement is 1% inspiration and 99% perspiration' is very true in engineering. Even if you regularly have wonderfully novel and fresh ideas, only the truly great engineers manage to make them happen. As demoralising as it sounds, this is the really hard part. You will repeatedly be told "no". You will encounter endless obstacles and the task will inevitably escalate into something far more difficult than you had anticipated. You must be bold to make such suggestions, but you must also have courage and faith to make the ideas a reality, as well as the persistence and discipline to follow them through to the end.

Realise also that the limitation of implementing improved and novel solutions is really mostly inside you. It will seem like everyone is out to get you. It will seem like everyone is opposed and the world conspires to terminate your idea. In the end, if you fail it is because *you* have given up and resigned.

It is because you have bought into others' opinions that it is "too hard" and "not really a great idea". The obstacles are guaranteed—how you deal with them is entirely up to you. Later down the track, you are much more likely to regret not stretching and challenging yourself as an engineer, than praising your conservative attitude. Furthermore, if you start on a small scale of projects, you will develop this approach into a habit. It will be virtually impossible for you to suggest a mediocre solution.

My final point with regards to originality is not to live in the shadow of anyone. If you have a father or mother who was extraordinarily successful, or your supervisor is extraordinarily successful, then it is unlikely that you will realise your full potential if you just duplicate their actions. The reason is that you have your own unique goals, talents and worldview. You will only have high levels of passion and commitment in those areas that are special to you. The only exception here is if you share the same goals and aspirations as the superior, which is unlikely. By all means utilise the experience of your superiors and allow them to pave the way for you (other people can be an important stepping stone to launch from), but at some point you will have to step out from behind their shadow if you want to achieve high levels of success.

Develop a Daily Ritual

Whether they like to admit it or not (and whether they realise it or not), the great engineers out there have a daily ritual. This is because the road is long, and in order to maintain your established, effective working habits, you have to repeat them over and over again. You can't let yourself fall to the mercy of your reasoning mind, which will only tell you to stop when it doesn't feel like maintaining the effort on a given day.

The ritual that you establish may be unique to you. It has to be a ritual that works for you. I also don't necessarily mean that it literally has to be daily, but you should repeat certain behaviours that you go through at least every week that keep you focused, on track and functioning with high effectiveness.

Take the Sunday night planning ritual that I mentioned at the start of the chapter as an example. Starting my week on a Sunday night, I will assemble a task list for the following week so that it is clear to me what I want to achieve. I make sure to develop this with positive emotions—with confidence and enthusiasm, and I make sure it is aligned with my grand plans. This way, on the Monday morning I leap out of bed with enthusiasm. When I arrive at work, I know exactly what I will be doing. Preparation is important, and if you've lived out your tasks previously in your head (it doesn't have to be on a weekend— Friday afternoons work well too), then nothing will surprise you when you get to work. As I also mentioned, I also tend to make a monthly plan ahead. That is, at the start of each month, I will note down the tasks that I want to achieve by the end of that month. You'd be surprised at how effective this is at keeping you on track and maintaining motivation, since it ensures that you are focusing on the right problems.

Spending an average of thirty minutes of time alone every day is important too. This is particularly important if you are a guy, as it lets you contemplate and reflect on events from the day. This could be behind a desk, or going for a hike or in the garage. If you are a girl, then the best type of self-reflection occurs through discussion of your daily events with your closest family members and friends. Although these thirty minutes will not always be possible, I would recommend making this a habit.

I would also recommend implementing a work-life balance schedule, even if you try adhering to it for only one month. I will talk about work-life balance in more detail in Chapter 6. For instance, outside of working hours, you may allow thirty minutes a day for self-improvement (reading some sort of engineering or scientific journals or perhaps leadership books), as well as thirty minutes for fitness. Needless to say, you should also allow some time on hobbies and relaxing in front of the television. The exact breakdown of these activities is obviously highly personal since only a certain formula will work for you optimally.

The vast majority of your time should be spent chasing your grand goals. There is no way to cheat the system. You can't lounge around and enjoy the spoils at the same time. The best breakdown that I have come across is the 90-10 rule. That is, for 90% of the time you should be chasing your goals while for 10% of the time you can be goofing off!

Since celebrating your successes is so important (I know I've reiterated this over and over again), then I will also make a note in my diary listing the successes of a day. For example, if I finished a book, or a paper, or won a grant, or sought advice from someone, or solved a problem—anything challenging or momentous (in other words, any milestone or achievement)—I will write it down in my diary as a dot point, and celebrate it in my mind with as many positive emotions as possible.

CHAPTER 4

Maintaining the Right Attitude

You may have noticed throughout this book that I place an emphasis on the internal aspects of engineering—that is, your values, goals and attitude. As a graduate this must seem puzzling. You probably expected the most revered engineers to be the technically superior ones—the most knowledgeable and fastest thinking. While technical ability is important, in the long run it is not as important as the right mindset.

Of course, being technically strong at what you do is very important. It is therefore critical to constantly improve at what you do. As you are faced with challenges and tackle them, your skills will improve. Challenges will make you better. Investing more time to hone your skills will also make you better. Attending training courses and being active in your community, and learning from the best will greatly improve your abilities. Your abilities are indeed important. *You are rewarded (that is, hired and paid) based on the service that you provide.* Regardless of what field of engineering you are in, the better you become the more you will be sought after and paid.

However, without the proper attitude, you will only reach a small proportion of your potential— regardless of how much work you invest. Conversely, with the right attitude almost any task is eventually achievable. Both aspects are required to be successful—being competent at what you do and having the right attitude—but I think that most engineers tend to falter on the latter. Having said that, I am not suggesting that you abandon your desires of becoming a technical guru in order to pursue some inner karma!

Early in your career, it is certainly acceptable to focus entirely on the external aspects of engineering. You need to build your technical skills and your experience in communications, dealing with people and industry standards. These are, early on in your career, your bargaining chips for any new job or promotion. Early in your career, you will naturally have more motivation for the technical aspects so certainly focus on them. You will definitely move forward more rapidly as an engineer the better you become. By all means, attend training courses, practise your skills, read books and so on.

The same applies to coming up to speed with the 'rules' of the workplace. Your organisation will no doubt have its own internal workings and policies and you need to become familiar with these. The same can be applied more generally to your industry. Time spent on this, particularly if it is of interest to you, is time well spent. You are learning and developing in a relevant field that will help you in the future.

However, you will inevitably reach a point when you realise that your true, big accomplishments haven't been the result of just technical skills, but your inner drive, enthusiasm, careful planning and general attitude to engineering. Similarly, you will also realise that most of your failures aren't caused by lack of technical skills, but with how you deal with problems. This seems marginally important at best when starting out, but becomes all important later in your career, as challenges become more and more complex. Furthermore, improvements to your technical skills will happen as a matter of course and in most cases it is not particularly difficult to improve technically. Any technical area can be improved upon with the right level of commitment and discipline. However, the wrong attitude is much more difficult to attain, but like anything else it is something that can developed. If you have both the internal and external aspects mastered, then you will be a rarity—a master engineer: perhaps only 5% of engineers fall into this category.

This chapter therefore presents what I regard as a healthy inner attitude to maintain as an engineer. This attitude will foster determination, perseverance, clear vision, planning of milestones and career goals, self-awareness, an awareness of others and what drives them, an ability to interact with others by building trust, general efficiency, and a healthy attitude towards your work and your peers. It will also support the healthy development of your technical skills, as well as your people skills. It will teach you how to get the most out of your career and how to be the best employee you can be. You will realise that you must treat engineering as a journey rather than a destination, and you will learn to enjoy that journey. Your new-found attitude will also warn you of the dangerous behaviours and mindsets that should be avoided, and trigger warning bells when you interact with people exhibiting these qualities.

Be Grateful

You should always be grateful and maintain an 'attitude of gratitude' every day. This is one of the healthiest attitudes you can possess, but it is actually harder than it seems. There may be times when you feel overwhelming gratitude, but it can be hard to maintain. The reason is that there is no easy measure of gratitude and how much you should feel entitled to. It is entirely a matter of your own perspective. This works in our favour since we can modify our point of view, even when we can't change the facts.

It is a case of 'is the glass half full or half empty?' You can look at any situation with different levels of gratitude. You may feel like you are the luckiest person alive doing what you do, having been through what you have to get there. Alternatively, you may feel quite the opposite. How you feel has little to do with reality. You may feel like you've been dealt bad cards—that you've been treated unfairly or that you deserve much more. No matter what your situation is, it is important to feel grateful and this should be a key aim. Perhaps you need to consider those less fortunate than you—those who haven't had the same privileges, such as the same education or the same support from family perhaps. Do you know of others who failed to land a job? They work for a poor company or have an annoying boss? Do you know of any new graduates who have had to quit engineering? Perhaps they didn't even get through their engineering degree at university? It could have been much worse! It is far too easy to think that you've had it hard, but this is only a case of perception and perspective. Unfortunately, the mind has a way of focusing on the negative aspects, which are usually insignificant, while forgetting about the positive aspects. *We are programmed to focus on problems*. The key is to be aware of this constantly and remind ourselves of the positives.

Exercise 15: Think of three aspects of your career that you have taken for granted but really should be much more appreciative of. For example: the ability to study, the ability to work with great people, the ability to have the freedom to look for the job you like, flexible hours and so on. Then, think of three lucky turns in your career, that you had no control over. Feel grateful for these moments when the planets seemed to align just for you!

Develop a Mindset of Appreciation

Ideally, you should maintain a feeling of appreciation throughout the entire working day. This is easily said and simple to understand but you need discipline to achieve it. I personally allocate five minutes of my time daily—on the train, on the way to work, or in my office—to think of three things that I am truly grateful for. Some days these things are the same, while other days they are completely new. For example, I may say "I am grateful for having that software licence that allows me to run simulations all day long!" or "I am grateful for a great team." Sometimes they are not necessarily work related. For example, you could say to yourself that you are grateful for a nice car, or a nice brother, or anything else. Some of these things you earned yourself. Some of them were just gifts. This doesn't matter. *At least once per day, picture yourself as being one of the luckiest people on the planet for being in your situation, no matter what your situation is!*

You should never feel like your employer or anyone else is out to get you. They aren't! Most of the time, people could not care less about what you do. It is amazing how many engineers I meet who believe that someone or some part of the system is there to hurt them and cause them harm. This dangerous mindset comes from the belief that you are at the centre of the universe. This is a natural belief to grow up with, since we are raised as children to believe that we are innately exceptional, but we are not. We only become exceptional by our actions, which at the time that we are children are yet to come!

If someone has done something that upsets you, then chances are that they didn't do it to upset you. They have their own interests in mind, as well as their own frustrations and often very personal problems that they are dealing with. They have an ego and they have their own issues and fears. People are complicated and far from perfect, but very few (if any) hate you! They have their own interests which shouldn't concern you. *You are responsible for yourself and no-one else. You cannot and should not try to control anyone but yourself.*

When you are convinced that someone is determined to

cause you harm, realise that they are probably dealing with their own inner demons. The actions you should take have been covered in the previous chapter. Generally I believe in an attitude of openness, as well as being firm and expecting mutual respect. You should not try to change anyone, but if their behaviour is unacceptable to you, you must make a stand or this behaviour will rear its head again and again, and at times from other people and directions! Of course, you should do this tactfully and professionally and never by throwing a tantrum! It will rarely be the case that someone is out to destroy you (and even if one in a million people indeed wish you harm, it does not help you to maintain this mindset for the vast majority of others).

Believe that People are on Your Side

People almost always have good intentions, and this is particularly true for engineers—you need to believe this. They are on your side. Your employer only wants the best for you. However, in the vast majority of cases your mind worries and tells you otherwise. The working world is simply a reflection of your attitude. If you have a negative attitude, then negativity is what you will see, and it will seem as if no-one is on your side. This is why it is imperative to have the right attitude and have faith that the working world is on your side!

Try to always maintain the belief that people are there to help you. This includes your managers, peers, clients and other associates. It is all too easy to fall into the trap of believing that some people are out to get you, so develop the self-awareness to monitor this fear when it inevitably creeps in. It is natural for our brain to think this way. The real problem with this fear is that if you develop it, you will actually act it out. That is, you will be reluctant to seek advice from those people when in need, since you won't think that they will be willing to help you. Sometimes, subconsciously and inadvertently, you may even do things to harm them, or refuse to help them because you harbour this ridiculous and unjustified belief. Once they sense that you are holding a grudge against them, they may be less inclined to work with you in the future.

The optimal mindset is to have faith that the vast majority of people out there would bend over backwards to help you. From my experience in the engineering world, this is not far from the truth. You will then have everyone's best interests in mind and you will be attentive to their needs. This is bound to come back to you. In fact, if you maintain this attitude over time—and as you build relationships in your career—you will reach a point where any problem is just a phone call away and everyone seems to jump at your needs! Building relationships with people is discussed in the next chapter, but it is important to realise at this stage that people are your greatest asset—they are not your enemy!

Be Humble

Always be humble. Realise that you are not at the centre of the universe—you are just a tiny piece of the engineering world. This doesn't mean that you should reel in your ambitions. Far from it. You should still aim for the sky and believe that anything is possible—because it is. The truly successful engineers also seem to appreciate the shortness and fragility of life. They realise that your very existence (not to mention your existence as an engineer) is precious, and can be taken away at any instant. What would you do if you only had one month left in engineering or on this planet? What about one week? What about 24 hours?

Being humble does not mean being shy. While it is normal to be shy (there are always going to be numerous personalities out there from the extremely extroverted to the extremely introverted but I have not seen evidence to indicate that one type of personality is superior in engineering), there will be times when you will have to get your message across and act loudly and confidently. Likewise, there will be times when you have to curtail your commentary and absorb information.

When someone outperforms you, give them credit. I don't just mean that you should congratulate them—this is up to you. But realise deep inside that they probably deserved their achievement. Don't instinctively attribute it to good luck or good fortune, a good education or good looks! It probably wasn't any of those things. Accept that they simply did a better job than you did. Their motivations and many of their actions are masked from you. You don't know exactly what they did and how they did it, but they got the job done. Again, for those one in a hundred cases where this is not the case, it is far healthier to maintain this attitude despite the unfortunate reality. *Live* this attitude. Focus on harnessing your own efforts, and not criticising the efforts of others.

Be Comfortable with Failure

I've covered the importance of making mistakes in Chapter 2, but I will reiterate what this means from a different perspective. What I mean here is that you shouldn't fear failure. As crazy as it sounds, you should see failure as your friend. As hard as it is to believe, without failure, you can't make progress as an engineer.

The key stepping stones in my engineering career have occurred just after massive failures. If you would like to achieve something big, don't be deterred by the possibility of failure. I'm not talking about cases where you are building a tower, and you know it will collapse—in this case you should definitely stop and reassess! I'm talking about situations where the ramifications of the failure are not catastrophic, and you have taken all reasonable and professional steps to give something a go—and you are mostly confident it will succeed. We tend to avoid any tasks involving uncertainty since a potential failure will make us feel uncomfortable, yet these are exactly the tasks that often offer the greatest rewards!

I've met countless engineers who are reluctant to apply for higher positions or new jobs altogether because they fear the possibility of being rejected. If you are not applying for a promotion because you fear not getting it, then this is not a reasonable excuse. If you are not selling your ideas and abilities because you fear that they will be rejected, then you need to carefully weigh up the cost of failure. What is the worst thing that could happen? Now think about the best thing that can happen? Is your decision really rational, or are you held back by fear? Where does your action fit in with your previously established value system? Are you a person who seeks to move the world forward with your ideas and actions, or are you intending to keep them inside?

The more that I've explored this concept of fear and the extent to which it holds us back, the more I've realised the importance of nurturing your confidence. This is what I will touch on next.

Foster Your Confidence

Confidence is like a tank of water that is gradually filled up with your successes. Throughout a healthy career (notwithstanding the usual ups and downs) this level should gradually be rising. As your skills improve, and you use these skills to accomplish more and more challenging tasks, your confidence will naturally rise.

Having said that, I have met new starters with extreme confidence, and in such cases this confidence can often take a hit. In my personal opinion, it is generally better to be over-confident, than to lack confidence (presuming that that you are adhering to the other principles outlined in this book related to treating others respectfully), or have problems in acquiring confidence. A confidence hit can take years to recover from (some people do not recover throughout their entire careers!) and your confidence can be sapped in an instant. To build your confidence back up can take time and effort.

I've also seen the 'fake it until you make it' philosophy widely applied. Personally, I think that if you can improve your confidence, even through slightly manipulative and artificial means, then do it—such is the importance of confidence! Once again, this is assuming that you do not break the other basic principles that I've outlined in this book. That is, you must maintain your professionalism and integrity while becoming confident at the same time—this may not leave a lot of room to move in terms of your behaviour, but it is possible!

Young engineers will often tell me when I ask them why they didn't ask for more responsibility or a promotion, that they didn't feel confident or adequate. The same applies when asking for a raise, though in this case you can actually cause some damage to your reputation (at least with your boss) if you don't say the right things. The key to moving yourself forward in terms of your confidence is to understand your massive potential and have faith in your unique skills and abilities.

Let's use the situation of asking for a promotion or raise as an example. In this situation you need to prepare—find out who is earning what, and determine exactly how your work history and skills match the new position. You want to be going into the boss's office armed with the right level of background knowledge. You also have to be armed with a list of your successes which will demonstrate that you are capable of rising up to the challenge. It is important to note, as I've said before, that there is no free lunch—you can't expect a raise or promotion doing exactly what you were doing before. You have to bring more to the table and you need to demonstrate how you will do it.

Your boss is more likely to give you a raise if they believe that your worth is higher than your remuneration, and that you could relatively easily land a job elsewhere. It is therefore essential that you know your self-worth and what the company and the market needs. Once you have this established, then together with the clarity of your goals you should have the confidence necessary to take action.

Most young engineers—particularly those with several years of experience—are often held back by their lack of confidence. They do not understand their potential, and they don't thoroughly assess the situation. They are often too complacent and don't believe they deserve (or could work towards) anything better or bigger. If it is just a hit to your ego, then get used to doing it. The attitude of "I will look like a fool" or "I don't want to be pushy" or "I don't deserve it" is mostly ludicrous. If you want it, then go after it. Don't be ashamed of what you stand for. Just make sure to be respectful when discussing and negotiating, as always. Senior managers know this, and this is why they are successful. They are often not much better than you technically or professionally (the quality of their design work or report writing may not be far ahead—in fact, it may be inferior!) but they know how and when to negotiate and they are not afraid to do so.

Focus on Giving and Adding Value

When I first heard this concept as a new starter, I thought it sounded ridiculous. But when I think back on my career, it was always the work I carried out without feeling a need for something in return (and with the utmost consideration of other people's needs) that resulted in the greatest success. Have a think about the most outstanding engineers and business people out there. They all achieved their successes by delivering a product that made other people's lives better.

It is a counterintuitive approach to take. Initial logic would suggest that you should do the very opposite. Most people will think, "What's in it for me? and "How can I maximise my return?" It certainly seems like the greedy approach will result in the greatest dividends. However, deep down inside, we know that this is not the case. We don't just steal the first nice car we see on the street, and we understand that investing in an education will generally pay dividends further down the road.

Our value as engineers is defined by, well, the value we add to our projects, the workplace, our company and the world! Therefore, if you focus each day on providing the most value you possibly can, then you will become a more valuable engineer! It sounds so simple, yet most people only see luck, politics and personal connections as the reasons for success, but this just isn't true.

Always focus on adding value, regardless of what your profession is. I rarely ever see this attitude applied day to day. For example, consider your time in a supermarket, cafe, or consider the products you use. It is often apparent that people could have easily done a better job, but they didn't! They didn't think it would matter. But it does! It's just that their reward doesn't get back to them directly and not quickly enough to warrant the additional effort. How many times have you been treated rudely? Sometimes the person is just having a bad day—they've broken up with someone, their car has been towed, or they don't like their job. Who knows! But you decide never to return there. You may leave a critical review or mention the business to friends. The person may have another bad day and the business eventually suffers. When the business is finally forced to act, they may look for new staff and that rude person will be first in the firing line.

One of the other reasons why this mentality works in the long run is that it allows for the natural reward mechanism of the universe and the associated lengthy time frames to play out naturally. Doing the opposite (focusing on obtaining a reward) only breeds a mentality of entitlement. This is not a problem when we are rewarded, on time, and as expected. However, this rarely occurs. Difficult projects almost always take longer than expected. This then leads to great frustration, and wanting even more! The feeling of entitlement snowballs, until *we begin to feel bitter for all our hard work, and we start to blame others and the 'system'.* We eventually become discouraged and seek to obtain more and more for ourselves or we resign from our efforts altogether.

I've talked about Receiving Your Reward in Chapter 2 but since it is an important element of this discussion I will reiterate it. You must be patient when awaiting your reward. It may come a year later, in a very unexpected form. You must have faith that effort spent in engineering will result in a reward from engineering. The engineering world can only be seen as a 'black box'. The operation of this reward mechanism is far too complicated to understand or model, so why bother trying? You must have faith that if you have done everything mostly right, and you are headed mostly in the right direction most of the time, then you will eventually be rewarded. This is why those ingredients for success of a plan (knowing where you are going), feedback mechanism (knowing when you are making progress) and changing direction when required are so important. They essentially guarantee that you will eventually be rewarded. They are three ingredients that should not be forgotten. There is a fourth ingredient which has also been discussed throughout Chapter 2 and at the beginning of Chapter 3—and that is perseverance.

The mentality of striving to add value is complementary to the mentality of giving, since you can do both at the same time! From my experience, everything that you give comes back at you multiplied. However, it must be truly selfless giving—you must not expect anything in return. Of course, there are some major caveats here. I'm not saying that you should go out into the street and give away everything you own! You must give in the context of your goals and aspirations that are clear to you. Being charitable or philanthropic is something different. Of course, if one of your goals is to be charitable then perhaps this would be an effective strategy. What I'm saying is that if you can easily solve a person's problem or offer a solution, then do it, without necessarily expecting anything additional in return.

Once you have mastered this mindset, then try to give *more* than is expected. I only see this characteristic in the very best employees. If you ask them to do something, they will not only do a fantastic job, but they will do something extra which you were not expecting, and they will do it with a smile. This really leaves a lasting impression.

It can be difficult to appreciate the notion of giving when it is not clear what people or what cause you are actually contributing to, since each time it may be someone or something else. Sure, you are working for your employer; however, you should think that you are contributing to the 'ether'—for the benefit of no-one or anyone, or perhaps all of mankind! As discussed above, your giving must be genuine. If it is not, and you are actually expecting something in return, then when you don't get it quickly, you will become frustrated and return to a more selfish attitude. Maintaining the desire to give also means that you give people the benefit of the doubt. What is the worst that can happen? They will receive a service from you for free. So what? If this bothers you then perhaps you need to redefine your value system. *You become more valuable as an engineer by what you give, not by what you take.*

Note also that expecting something in return reduces your feelings of gratitude (its importance was discussed at the beginning of this chapter). Rather than feeling grateful for the situation that you are in, you feel jaded. Do you think that you will be likely to press on? What will your attitude be towards your peers and supervisors? What about your workplace? Will you be working harder now? Accept and have faith in the premise that you will be rewarded on the contributions that you make. Focus on giving and adding value in everything you do.

Finally, be generous. If someone asks for thirty minutes, give them an hour. Think about it from their perspective. How would you feel if you approached someone for a favour and they gave you more than you expected, and with a smile? Or if you take someone out for a drink, try shouting them once in a while. It makes little difference to you (unless you are really under the pump financially) while small actions like this can leave a lasting impression and signal your generosity. Now, I'm not saying that you should pay for everyone indiscriminately, or even pay for anyone. What I'm saying is that when there is an opportunity to be generous, then do not hesitate. The more you give to the engineering community the more the engineering world will reward you.

How to Stay Positive

Everyone has times when they feel on top of the world and times when they feel down and dejected. This is natural. Sometimes projects fail, or your computer breaks down, or you have a conflict with someone, or perhaps your career takes a hit. Maybe you even lost your job, or failed some subjects at university. There is no preventing all of the undesirable events that are going to happen to you, regardless of whether they are in your control or not.

As described earlier, failure is an integral part of engineering. You must have faith that all failures happen for a reason, even though the reason may not be apparent to you at the time. Sometimes it can take many years to figure out why something didn't turn out right, but when you do, everything makes perfect sense. For example, you may have lost a contract to another company, only to learn several years later that the other company was far more experienced.

Not only does the normal, productive work week bring failures your way, your perception is a significant variable too. If you are in a good mood, you perceive things differently to when you are in a bad mood. For example, if you have just received a reward, if you have an adventurous weekend to look forward to, if others seem to be treating you exceptionally nicely, or if you are simply in a peaceful state, then your perception of the world will generally be positive.

From my experience, it is often the perception and not the facts that is far more important. Sometimes you see people who are in shocking situations—either in terms of their careers or personal lives, yet they are happy. Conversely, others may seem to have it all (a role you aspire to—generous funding, prosperous contracts and so on) yet they can be miserable. Hence, how you look at a situation is far more important than the reality. This is another important point to grasp when reacting to undesirable outcomes. I am not saying you should be happy when disaster strikes, but realise that in the vast majority of cases the situation is not as dire or prevailing as you may think. Learn, move on and stay positive.

Most importantly, you cannot allow those demoralising moments to stop you in your tracks. Successful engineers know they are not always going to feel positive about their work or themselves, but they plough on anyway. They do not allow negative feelings to jade them. They are able to tame and control their emotions and see the big picture. They know that a positive experience is just around the corner. Going through this repetitive process will help you. Once you've experienced enough ups and downs, you will learn that there is always going to be a cycle.

Don't be fooled into thinking that some people are always stuck in a productive part of the cycle, or they are exempt from this cycle. This is impossible. Even if they could control their perceptions entirely, external events are always going to be unpredictable. This is the nature of engineering, and life, for that matter. This is why our brains are structured to be so versatile. In fact, from my experience, the people who appear to always be positive are usually those who have been hit the hardest. They have survived and they've subsequently learned how to adapt and deal with hardships. They don't have it as easy as you may be led to believe.

It is also interesting to observe how a positive mentality

and outlook influences others. I'm sure you have noticed how people seem to react differently to you when you are in a good mood. They feel and draw from this energy. They become more forthcoming, helpful and more likely to share their ideas with you. You will get the best out of people by being in a good mood yourself. Other people can be a mirror that reflects your mood.

Staying positive, needless to say, has numerous benefits. You tend to be more relaxed when you are positive which means that you are more creative. You also tend to be more driven and more focused.

This is another reason why it is important, whether in your professional or personal life, to be good at something—or to have something that you enjoy doing. It is then useful to revisit this skill during those times that you are not feeling confident. It may be an area of study or a project—something that you are good at or have done well at in the past. You can also use this skill to draw from if you become dejected. It will help return your mind to a positive state and renew your composure and confidence.

Exercise 16: Think of three talents, skills or events that you are proud of. They don't have to be work related. For instance, your ability to solve a certain kind of problem, or having your team win a football premiership, or winning first prize in a painting competition. Enjoy the feeling of confidence that they bring. Revisit these events when you need a confidence boost.

You may think that you are in a catastrophic situation, but there is always a positive angle that you can take. In other words, there are always two sides to every coin, even if it is not apparent to you. Hardships with no apparent rewards often end up being worth their weight in gold.

As I've said before, it may be useful for you when you are down to think of people who have it worse than you, or scenarios that could have turned out worse than they actually did. People often live with this idea that they are failing at everything, and their situation is dire. This is usually far from the truth. It is simply human nature not to linger on the positive aspects, while seeking out and worrying about the negative aspects. Worse still, the mind can be engulfed in worries that haven't yet happened, and have little realistic chance of ever happening. Such is the often cruel inner workings of the human mind! This leads on to the next section of the mind and mindfulness.

Mindfulness

When a colleague mentioned mindfulness meditation to me when I was a new starter, my initial reaction was, "What a waste of time! I'm too busy for this airy-fairy nonsense!" I'm sure you didn't think you'd be hearing about mindfulness in an engineering book. It is highly likely that you feel the same way that I did back then. It may be very true that you don't need mindfulness right now, as I didn't back then, in which case feel free to skip this section. However, I can guarantee that there will come a time when mindfulness will potentially play an important role in your engineering career and general well-being, regardless of your disposition, your state of mind and your character. There is a very simple reason for this—we all share the same primitive brain structure and it did not evolve for the reasons that it is being used for today—particularly not for dealing with an onslaught of engineering problems day to day.

Although you may be at peace with yourself now, there will come a time when you have many problems to assess and keep track of. You will have competing priorities, and you will be dealing with people who may not see eye to eye. You will be in high-pressure situations with no obvious escape route. You may be doubting yourself, and endlessly digging up events from the past that are really not relevant today. Most people also tend to worry unnecessarily about all the future problems that *could* happen.

Mindfulness meditation is the process of consciously moving your mind into a peaceful and relaxed state so that it can perform at its peak. Of course, some level of concern about past events and future events may be healthy. However, we tend to do it way too much. Most available resources on mindfulness talk about being in the 'present'—living in the moment and not thinking about the past or the future. It is well accepted that our tendency to linger in the past and worry unnecessarily about the future is detrimental to our success. In some case, this stress results in anxiety disorders and depression. This is remarkably common, yet the situation is created entirely by ourselves. There is no actual threat to us, either in the past or in the future, yet we create one that hampers our lives!

Being at peace, which means not having worries or concerns gnawing away at you—either from the past or the future—results in you being more relaxed and more creative. Your relationships with people improve since you feel good about yourself, and you are generally happier. Yet very few people realise that achieving this state is something that we can actually control. Most people don't see the mind as a precious resource of ours, that needs to be nurtured and maintained.

The evolution of our mind has for the last several decades fuelled research and speculation. The cognitive mind separates us from other species. Interestingly, it is not homogeneous in its structure. There are numerous parts to the brain that are split in two hemispheres. Some workings are understood today while others are not. The important point to note here is that the evolution of the brain has taken on a particularly interesting path as compared to the evolution of most of our other organs. Our brain is made up of different sections that have been tacked on as the evolutionary need has arisen. We can't be sure what these evolutionary needs were—they may have been survival, creativity, or relationships. My point is that because of the immense evolutionary time frame involved in forming an intelligent mind (as opposed to single-cell life) the brain evolved in sections, like different-coloured chunks of play dough being stuck together. These sections then communicate with each other through electrical impulses. We roughly understand what the function of each section is, but we rather poorly understand the combined workings of the brain. The key point here is that because of the unique evolutionary path in creating the brain there are undesirable by-products that we all must deal with.

Most notably, your brain is naturally in a constant battle with itself. Since the brain is comprised of modern cognitive sections and older autonomic sections, in left and right hemispheres, then these internal *arguments* are destined to occur since the viewpoints from each section are rarely consistent. No part of your brain is *you* since your brain is not *one*. Every high achiever has achieved some level of self-awareness and is aware of the limitations of their brain and takes action to work with it, and not against it.

If you have a tendency to dwell on past events, or worry about future events excessively, and the worry is getting in the way of or interfering with your day-to-day work, then mindfulness practices may be of benefit to you. The key to achieving mindfulness is to realise that the workings of the brain are natural and unavoidable. The solution is to establish a third-party view of the situation. That is, realise what it is that your brain is doing and 'observe' it, but don't engage it. Engaging your negative thoughts—by analysing, justifying, or applying logic—will only fuel them.

If you have a tendency to dwell on the past, then realise that history is never how you remember it. It is easy to project what happened into the current day, but our memory is notoriously subjective in how it remembers events. It stores a distorted image of the past that is rarely accurate and never objective. It can't be relied upon for an accurate record of the past. Secondly, realise that we are constantly changing. Therefore, we were very different back then, and the situation would have been perceived differently. When we dwell on the past, our minds don't compensate for how we have changed. It is therefore not appropriate to project our *new* self into the past, since we were so different.

If you have a tendency to worry about the future, then realise that such worries are almost never justified. When we worry about something not turning out right, our concerns are invariably magnified out of proportion. Not only is the event we fear far less likely than what we imagine, but the consequences are also far less severe. We also tend to drastically underestimate our abilities to respond should something unfortunate occur. I'm sure you've had this experience in the past. Perhaps you were worried about giving a presentation, but in the end the presentation wasn't as bad as you feared it would be.

Of particular importance to engineering is being able to convert a negative state of mind to a positive one. Everyone has moments when they have a positive state of mind, but it is what you do with your negative thoughts that matters. This is where strong skills at mindfulness meditation can help drastically. It will be a valuable tool in your arsenal as a high-flying engineer if you manage to master your medieval brain!

There are some established routines for mindfulness meditation, and I suggest that you pursue some books on the topic and establish your own routine if you think it will help you. Realise that it takes time and practice to achieve high levels of mindfulness but do not underestimate how powerful these routines can be. Personally, I have found that the most effective routine is a simple, daily nine-minute appreciation exercise (I highly recommend the book *Awaken the Giant Within* by Tony Robbins where a variation of this routine appears). For the first three minutes, focus on aspects of your life that you are truly grateful for. You can think of your girlfriend, or your family, or your car, or the nice meal you just had. Spend a minute on this, and then move on to thinking about something that has happened that seemed fortuitous and out of your control. For example, an old friend called you out of nowhere and you had a great night catching up. Then, for the third minute, think of something great that has happened to you that you have earned. For example, you worked hard on your degree and then graduated! At the end of these three minutes you should be feeling a deep-down appreciation and inner peace. You should feel true gratitude. You will be amazed to find that it is possible to achieve this state in just a few minutes.

Then for the next three minutes, think of three individuals or groups and wish them well. For example, you may have an auntie who has a cold. Or perhaps there is someone you don't like—this could be a great opportunity to feel empathy and finally resolve your inner conflict.

Don't allow your mind to drift from these thoughts and the thought of your breath. If it drifts towards your problems or anxieties, gently coax it back. Nine minutes of such basic mindfulness meditation techniques should really not be hard to achieve, even as a new starter. If a problem enters your mind throughout this time, try to visualise it as passing straight through—don't engage it. Picture yourself as being untouchable—nothing can shake you!

Finally, note that the ultimate aim of mindfulness for engineers is replacing your negative worries with some ambitions, which are essentially healthy worries. Your mind must be occupied by something throughout the day, and unless you give it something worthwhile to digest, it will just dig up events from the past and unlikely problems from the future. It will become like a trashy talk show. Don't let the quality of your thoughts decline, or you will start living your thoughts. Therefore, for your final three minutes of mindfulness meditation, think of your ambitious and exciting plans for the coming day, or for the coming week or year! Think about them as if they have been achieved. Perhaps you are hoping to finish a design or a report, or perhaps you are saving for a car!

I normally do this nine-minute exercise (which is summarised below) at the start of every day. When you are done you should feel cleansed, fresh and ready for another day. Of course, feel free to modify this routine depending on what works for you, but I would recommend establishing some basic habit of mindfulness practice. I believe that simple mindfulness activities are a long-term investment which lead to a healthy mindset in the long term. A healthy state of mind can have a drastic effect on your work quality and your general well-being.

Exercise 17: For three minutes each, focus on the following three mindfulness meditations:

1. Think of three things that you are extremely grateful for: something grand, something fortuitous and something you have earned.

2. Genuinely and intensely wish one or more people well, as if you could influence their situation (for example, a colleague's success or a family member's well-being).

3. Think of your grandest goals and see yourself achieving them with clarity and excitement.

As a final note for the mindfulness section, beware of the desire to detach from reality—to run and hide. If you do find that your mind is digging up problems, it is important not to run away from them, but to face them (that is, accept them), since they won't go away until you do. Examples of people running away from problems include excessive exercise, shopping, hobbies and drinking; becoming emotionally detached from tasks and people; and avoiding certain scenarios and situations.

The Ego

Everyone knows what a strong ego is, but it may come as a surprise to you to see it mentioned in a book on engineering. This is because it can play a damaging role for engineers and it is so widespread. Once you understand this you will be impartial, and when you see others being driven by their egos, you will understand their emotions. You will be less likely to take their comments personally and you will make balanced judgements.

The ego is something that we construct—a set of rules to live by. These rules tell us how we should be, and how the world should be. The reason why this is a problem is that these rules are a creation of our minds, and may not represent reality. For example, you could be saying to yourself:

- "I'm the best engineer there is!"
- "I'm more intelligent than all of these people. I can't possibly learn a thing from them."
- "I know much more than him."
- "I am more valuable than her."

- "My father/mother/brother/sister was an expert on this, so I am an expert too."
- "I own this branch/project/product."

The ego often covers up insecurities. The real problem is that *the ego prevents us from seeing ourselves as we really are, and prevents us from taking the action required to change and improve.*

Often people have large egos and when their results don't turn out as they expected, their egos are shattered and they lose control. Anyone who has ever had reality crush their ego knows how painful this is. Eventually they reinstate their egos, and give themselves an explanation of why they failed. It is important to question your ego when it is crushed. The working world has a funny way of exposing your ego. I've held on to strange and ambitious claims only to have them exposed in unexpected and embarrassing ways. It is a wake-up call to have your ego called out.

When your ego takes hold of you, then you get worked up. You may argue, sweat and even turn red in the face to defend your position! Afterwards you may become aware of your unusually aggressive reaction and wonder what on earth triggered such a response! It is simply because the ego is so strong and so deeply ingrained that we can't help but react when it is challenged. Breaking down the ego is not easy, particularly if it has been built by years and years of reinforcement. Breaking down the ego does not mean ignoring the ego, or ignoring other people's comments that challenge the ego. Eventually there will come a time when it comes out! The right approach is challenging our own ego, recognising the ego and asking ourselves where it came from. Questioning the ego, and seeing reality is the key.

Your ego can be strengthened by other people. You then take on this false image of yourself and defend it with your life. It is fine to be praised and rewarded for a job well done, but it is important to be grounded to reality. Just because you work for a high-profile employer, or on a high-profile project, or you've just attained a great qualification does not make you a superior person to anyone else around you.

There are a couple of caveats here that I should mention. Of course, as I've explained many times before, all achievements should be celebrated and enjoyed since they are taking you somewhere (hopefully in the right direction!). It is of critical importance to appreciate your achievements—you need to give yourself an inner reward so you have a closed feedback loop. You need to feel good about yourself. Also, I'm sure that you are improving every day, and you probably have numerous achievements under your belt. However, each day is a new day which effectively represents a *reset* condition. You need to be in an optimal, hungry and learner state every day to consolidate your progress. Unless you maintain the mindset that got you here, you will stop making progress! Therefore, your successes and your status should not change your relationships with other people, and they should not diminish your desires or the realisation that anything great has to be worked towards.

As hard as it may be to accept, realise that you probably have a dominating ego at times. The harder you work to address your ego, and the more neutral you become, then your yardstick for measuring other people's egos becomes more accurate. More importantly, your ability to see the truth in each situation will improve. You will see an ego-driven opinion response with much more ease. You will tend to respond in a balanced way free from emotion. It won't bother you anymore if someone challenges you—you won't feel angry as you won't have an ego that is being challenged!

Consider the common situation where all parties at a meeting possess a strong ego. An animated conversation or argument is often simply a battle of egos and not the real people and certainly not the facts! Just perceptions! It is as if we were actors in a play. Furthermore, your perception of another person's stance or facade is tainted by your own stance or facade! This makes reality a far-removed concept in an ego-driven world!

When you eventually recognise and manage your ego, very few things will shake you or drive you into an emotional state. Criticism from others will seem to pass straight through you. You won't engage them, regardless of whether you agree or disagree. You will be at peace, relaxed and focused on what it is you want to achieve. When you are relaxed and focused, this is when your truly great ideas are likely to come—not when you are stressing about what others have said!

Stop Looking at Others

Stop looking at and judging others. This is just your ego at work—looking for ways in which it is being challenged or it may be challenged in the future. Do you ever say the following to yourself?

- "Is she better than me?"
- "What is he scheming?"
- "Why didn't I get that contract?"
- "Are they trying to undermine me?"

These thoughts are usually ego-fuelled, and usually unhealthy. You should be focusing on yourself and achieving your own goals, and not worrying about what others are doing. As mentioned before, you rarely have the information to judge anyone else properly, so why compare yourself to others? Secondly, comparing yourself without the facts leads to envy and jealousy. Once you are jealous of your colleagues, you stop acting in their best interests and in the best interests of the team, and they will stop acting in your best interests.

Note that while confidence is very important, it is different to ego-driven arrogance. Confidence is precious and takes years to earn—it gives you the firepower to pursue your ideas humbly and respectfully. Ego-driven arrogance is not precious—it is common and detrimental to you and your peers.

It is important to face up to and eventually let go of your ego. You can still thrive as an engineer with an ego of course (most people do), but when you develop a self-awareness of your ego, then you will be in a distinguished minority group. Your negotiation abilities and interpersonal skills will improve substantially, which will help you throughout your career.

Recognise that you possess an ego that is likely to run rampant unless you spend time looking at it, thinking about it and addressing it. You can often help egos from developing by rewarding yourself properly. In fact, this is one of the reasons why egos develop—people don't feel adequately rewarded for what they have done. They begin to judge others and determine that they have been hard done by. Focus on your little achievements; keep in mind the complicated reward mechanism of the universe, and don't compare yourself to others. These are the same simple rules repeating in numerous sections of this book, again and again. Finally, it helps greatly if you can see reality for what it is, which is the topic of the next section.

Seeing Reality

See things for what they are, and not for how you want them to be, or how others want them or how others see them. Being able to see things for what they are, seeing through all the noise, the politics and the fog created by other people's egos and beliefs will be a powerful tool that will serve you well throughout your career. It will take time to master. Over time you should develop an impartial, external viewpoint. I've heard many times before that events are never as good nor as bad as we perceive them, and I think there is a lot of truth to this. Reality normally lurks somewhere in between. You should strive for an external viewpoint that is neutral, impartial and uncoloured. Ask yourself: what would a reasonable, well-informed, insightful and clever, detached third party see?

If you do everything mostly right in your career, then each day you will be only slightly better than the day before. Regardless of what rewards you may receive, realise that progress is a slow and lengthy process. You don't suddenly morph into someone else when you receive a prestigious award or a diploma! As previously explained, it is the reward mechanism that is complicated and unpredictable, and this tends to pump up our egos (and sometimes shatter us when they are crushed).

Always be honest with yourself. It is very easy to subconsciously deceive ourselves for protection. We want to protect our egos, and this leads us down the path of creating lies when dealing with others and worst still, when explaining events to ourselves! For example, many of my colleagues have, like myself, applied for promotion opportunities. Needless to say, when applying for a position, most of your attempts are unlikely to be successful. It is largely a matter of numbers. After failing at a job application, I've heard people say:

- "I didn't want it anyway!"
- "Management positions just aren't for me, so I'm glad I didn't get it."
- "I heard that the successful candidate hates the job."

Of course, they are just fooling themselves. I've even heard people deny that they applied for a higher position after they found out they weren't successful. Their egos could just not handle the battering! The best line of thinking in this case is: "I tried to get the advance because it sounded like a big step forward. Unfortunately, I wasn't successful. I will try again next time and in the meantime press on with my busy work schedule." This is simple, honest and nothing to be ashamed of, yet only a small fraction of engineers have this mindset.

Finally, don't be selfish. Being selfish is another way of

feeding the ego. The need to be selfish means that you have an ego that is demanding to be fed! Furthermore, it shows that you are not prepared to see reality for what it is. You may see a glimpse of reality, but you really don't like the look of it so your ego wants to create an alternative world!

CHAPTER 5

The Importance of Other People

The further you progress in your engineering career, the more you will appreciate the importance of other people. As a generalisation, just about anything that you want in your life can be achieved through your relationships with other people. Your relationships are the single most important aspect of your working life (and probably your personal life). I expect that if you truly absorbed and appreciated this statement, then your life would be turned upside down right now! I know mine was when I learnt this.

The Introverted Engineer

This lesson is all the more shocking if you are an introvert, since it means you are eventually going to have to master your people skills. Furthermore, this skill can't be acquired through a two-day training course—it is a skill that you will continue to develop throughout your entire working career. If you are a socially balanced extrovert who likes to engage with people, then this realisation probably won't bother you—in fact, it will probably excite you. This is an area where you will have a strong head start on your peers. However, there are many aspects to relationships, such as developing self-awareness and listening skills that you can almost certainly improve upon.

If you are an introvert, this means that you will need to put on an extrovert hat from time to time. That is, for short periods of time, you are going to have to work in a zone that is very uncomfortable and tiring for you. It could be a meeting, it could be an interaction with a client, or it could be a talk or presentation. You will have to develop the ability to enter extrovert land. You will quickly learn that this can be tiring—exactly how tiring it is will depend on how much of an introvert you are. You may be able to survive one hour, one day, or maybe one week. The consequences will be that you will feel worn out at the end of the day, since social interaction for introverts is draining. The good news is that with time it won't bother you as much, and you will become just as good at relationship building and social interaction as any extrovert.

If you are an introvert, be aware that there is an upside to your introverted nature. You probably like spending time in deep thought. The level and depth of your thinking will mean that you will often come up with incredible solutions to problems. You will also possess good listening and analytical skills. That is, in meetings you will tend to absorb all available information before voicing your opinion. The important point to remember is that no matter how introverted you are, don't feel that extroverts have an upper hand on you. There are always two sides to a coin, and as you will quickly learn, transitioning from one personality type to the other, depending on what the situation demands, is not difficult.

Temporarily wearing the extrovert's hat is realistically unavoidable if you want to succeed in engineering. You will need to get out of your comfort zone, talk to strangers and speak up in meetings when every part of your body is telling you it is a bad idea, and you are coming up with every excuse in the world not to do so. Don't worry—this is natural and the more times you do it, and the more of a pressing reason you give yourself to do so, the easier it will be.

One major inhibition to overcome is a feeling of fear. Realise that this feeling is internal. It is a product of your genes and the way that you were raised, and it is not necessarily appropriate for engineering. What is the worst thing that can happen to you when you speak up? You may think that you will sound like a fool if you say something stupid. Realise that if you are on your path, striving towards what you want to achieve, then you shouldn't care about what others think! Another major inhibition is the feeling that your opinion hasn't been properly formulated. You are therefore forced to act out of character. Once you become familiar with this feeling and experience it enough times, you will feel comfortable about speaking quickly, confidently and assertively, even if you occasionally make a fool of yourself.

Realise that the simple act of active interaction and conversation builds rapport, regardless of what you have to say. Even if you say something that is not entirely correct, people will understand where you are coming from. As crazy as it sounds, people—at least emotionally—respond to your intentions (which are translated through other forms of communication such as body language) rather than the meaning of your words. Have you ever heard a politician talk nonsense for hours straight, yet his audience is still glued to every word? In any communication, it is often better just to keep the communication channel open rather than freezing up. When you freeze up or hold back, no-one knows what you are thinking and silence breeds feelings of contempt (based on other people's internal negativity, as explained earlier in the book).

The fact that you are reading this book implies that you want to improve and you probably already possess strong self-awareness. You are probably already one of the best qualified people to be asking questions in the meeting room. You would be surprised to learn what levels of stupidity are possible from those that you regard as the most capable people, so relax! However, don't forget those points I made earlier about respecting others—do not allow yourself to be drawn into childish arguments or battles of egos, and never be condescending to anyone. Speaking up and being impolite (for example, by speaking over the top of others or being inconsiderate) are two different behaviours.

Realise that your challenges interacting with other people are not unique. Perhaps you struggle with interacting with people, but others struggle with basic mathematics. Chances are they are also facing challenges on the relationships front—just about every senior engineer does. Do not feel like you are different or have been dealt a bad set of cards—everyone feels this in one way or another at some stage. The worst thing you can do is allow these feelings (which are essentially feelings of fear) to make excuses for yourself, or allow others to make excuses for you. Realise that everyone has challenges to overcome when it comes to dealing with people. Too often I hear introverts saying that it feels unnatural and therefore they don't want to do it. *However, if you are going to make progress and end up in a position different to where you currently are, you have to get used to doing things that feel different!*

Work on your social skills. I generally don't believe in working on one's weaknesses excessively since I feel like you should be building on your strengths, but your relationships with people are an essential part of your life—like eating and breathing. There will come a time when you are hampered by your ability to form and maintain strong working relationships with other people. Therefore, discipline yourself to mingle and talk to others—build networks, interact and speak up as much as possible. Allocate an hour in your week where you do nothing but meet new people and mingle. Most engineers sell themselves short significantly, and it's often because of their introverted nature. Be confident and don't be afraid to be at centre stage. Sit at the front row at meetings if a topic interests you, even if there are managers there. Be interested and absorbed in everything and listen carefully. Demonstrate that you regard yourself highly—that you are capable and worthwhile—and you will be treated accordingly.

What Other People can Offer You

The vast majority of new engineers don't appreciate how valuable others can be throughout their careers. Given the choice, they would rather resign themselves to a life of professional solitude rather than building strong networks and relying on others regularly.

Other people can be:

- A source of great ideas
- A source of help when you are struggling with some particular piece of work
- A source of tuition and learning. This is particularly important early in your career
- A way of opening doors to new projects
- A way of connecting with more people
- A valuable way of promoting your work and abilities. The vast majority of promotions and new jobs are created via word of mouth
- A source of advice
- Someone to hear you out and absorb your frustrations
- Someone to congratulate you on a job well done and give you that all-important positive feedback

For these reasons, you should treat your relationships with other people very seriously. In particular, if you have a good mentor to learn from, you can learn and progress in your field extremely quickly. I have seen graduates who were left alone with textbooks from uninterested supervisors, and others who were nurtured by caring supervisors who were able to convey their many years of experience. The difference in their development is unbelievable. This is one reason why as a young graduate applying for a new job, you should pay particular attention to your prospective management chain and their level of involvement with junior staff (the best approach here is to ask others already working in the area).

You should not be afraid to rely on others. Having others to depend on is essential, even early on in your career. Many young engineers feel they need to prove themselves by doing everything themselves, but this is not an effective strategy in the long term. Realise that most accomplished engineers have gotten to where they are through massive input of others, which they were able to instigate and harness. Of course, I am not saying that they didn't do the work, and that you should ride off the achievements of others—not at all. What I am saying is that, particularly early in your career, you need involvement from the best people on all facets of your work, so that you can learn and carry out your own work in the best possible way.

How to Talk to Others

You should aim to achieve a level of confidence with your communication so that you have no inhibitions in starting complex discussions with other co-workers, whether internal or external to your company, or asking for advice and favours. You want to be able to get up and confidently present in front of a roomful of people. You want to feel comfortable in taking a front-row seat in a meeting and feeling like you deserve to be there.

In achieving this, there is bound to be an element of 'fake it until you make it' which I have previously discussed in Chapter 4, but there is nothing wrong with this. In other words, if you vividly see yourself as a socially balanced, talented and successful engineer, then you will become one. Don't under-estimate the power of this effect, even if it sounds ridiculous right now. You will act out who you want to be and others will eventually see you as this person. Of course, with such an ap-proach there is a risk of appearing insincere and arrogant. I see this with young engineers but to a large extent this is acceptable since you are in your learning phase. It's better to do this now than to have to worry about who you are later in your career!

To improve at talking to people, you need as much practice as possible which means that you should be communicating as much as possible, in as many forms as possible, and going where it is uncomfortable. I generally don't believe in doing what isn't absolutely necessary. That is, I wouldn't seek out difficult conversations for the sake of having difficult conversa-tions, but there are many other options open to you for practice. Try to find something that is especially meaningful to you. You may wish to talk to your supervisor about getting involved in more 'people' roles and responsibilities. For example, volun-teer to chair a meeting every now and then (taking minutes for the first few meetings may be a good starting point to increase your confidence). Offer to give a presentation now and then on either your work or perhaps an interest outside of work (I've had a colleague give a presentation about his trip to Europe!). Hold a teleconference. Run the local social club or coffee club. Go and ask someone for a favour. Offer someone a creative so-lution. Practise giving compliments. Sincere compliments are not as easy to give as it sounds. Have a chat to someone you don't know about career advice. If it is someone you aspire to, they will usually very gladly give you advice. The more times you are placed in uncomfortable situations, the more quickly you will start to feel comfortable. I used to dread giving presentations and talking to management, but having been through both so many times, I now don't give it a second thought.

Your self-esteem when talking to others can be boosted by having a plan and being clear on what it is you want to achieve. Your plans will be challenged again and again, directly and indirectly, but if you have a clear plan combined with a compelling reason for achieving it, you will stand your ground. You also gain confidence from doing something well. This is why you should always endeavour to improve at what it is you do!

The alternative to accepting that relationships are important is resigning yourself to a professional life of solitude. I've seen this countless times before. Ironically, it is often the best staff from a technical perspective that do this. Every staff member tells themselves something to take them out of the game. No-one sees it as a major problem, even though it has drastic consequences for their careers. Examples of excuses that people give themselves are:

- I don't want a position that requires people skills.
- I'd like to focus on my technical skills.
- I'm technically brilliant and nothing else matters.
- Other people talk too much.
- Other people don't know very much.
- Other people just slow me down.

These staff members then slave away in their own offices, minimising their contact with others. Of course, there is nothing wrong with pursuing a deep-level technical calling. However, it is making excuses and lying to yourself that leads to problems, because eventually reality will catch up with you. You will be missing out on the many ways that other people can complement your own skills. Furthermore, by doing this you are failing to realise that in order to make progress you have to develop and you have to step outside of your comfort zone occasionally!

What are You Afraid of?

If you are still having trouble deciding whether to adopt a more people-focused approach to engineering then consider, "What is the worst thing that can happen?" Will you look like a fool? Will you feel uncomfortable? Are you afraid of hurting your reputation? Will you offend someone? Will you lose your job?

Firstly, realise the importance of operating in good faith. That is, as I've said many times before, the world and other people are not out to hurt you. If you don't purposefully try to deceive or harm others, then you should be conducting your work under a clear conscience. Of course, I am not saying that injustices don't happen—they do! However, in the world of engineering and in the long run, you will generally get what you deserve. Even if you are not convinced of this, then this belief is healthy for your progress. You cannot fear something that is not justified and not deserved—otherwise we would never get anything done! Too often I hear engineers explain their reluctance to take action (and I'm not just talking about in relation to dealing with other people) with a series of 'what-ifs'. They will ask, "What if he thinks I'm stupid?" or "What if she thinks that I'm trying to steal her ideas?" or "What if he doesn't have time?" Through this line of thinking, they are already rejecting their own suggestions before they make them! This is self-sabotaging behaviour, whether they realise it or not.

If you are genuinely working in the best interests of people (and by abiding by the content in this book, you would already know how important it is to respect others and give to others) then you should already have a clean conscience and good karma! In this case, you should learn to accept all of the consequences of your actions! This is such a key message. You can't be inhibited in taking action through the fear of consequences, if your actions are directly aligned with your goals which are based on solid principles.

Technical Versus People Skills

When you are starting out, you believe that your own technical skills matter most, and this may be initially the case. There is nothing wrong with always developing your own skills and expanding your knowledge. In fact, earlier on in your career you will naturally have much more of an interest in the technical aspects to your job. However, you will soon realise that there is only a limited amount that you can do on your own. As you progress further into specialist areas, you will quickly realise how little you know. Even within the fields that you regard as home, you will learn that there are people more experienced than you, and unless you commit yourself fully to a particular field, you are unlikely to be a world leader. I am not saying that you shouldn't be committing yourself to the technical aspects of your job—I believe it is paramount that you become highly skilled in your job—but realise that even if you become extremely competent in a field, you will only be marginally competent in other fields where you will need help. On top of that, you will also quickly realise that there is only a limited quantity of work that you can carry out each day. Your work can be easily magnified thousandfold through the use of other people. But there is much more to the importance of people than leveraging off and leading others to achieve your goals!

Virtually all of your good ideas and fruitful projects will be in one way or another initiated by something that someone else has said. Do not underestimate the amount and diversity of knowledge and ideas that are out there. Virtually any question that you have in your life, or desire can be answered by others. Chances are, other people have had the same questions and have already answered them. *The problem is that we are unaware and unappreciative of how much knowledge and inspiration is accumulated within other people.*

Other people's wisdom is an important key to taking massive shortcuts. As I have mentioned in Chapter 3, you can save yourself a decade or more of effort by learning from others. Other people hold the doors to your further progress. If you do everything right, these doors will open and your progress in engineering will be rapid and smooth. Do not underestimate how much of an influence other people have on your progress and success in engineering. While your skills and abilities form the foundations of your success, it is your ability to leverage off and learn from others, and relate to others and maintain a high level of respect that will bring the big success.

People skills are an important tool in your engineering career. *Most of your challenges and problems will be solved in conjunction with other people in a team, and not as a lone entity.* This can be hard to appreciate in the early stages of an engineer's career when they think they can change the world and succeed alone. Your skills in communication, working together, negotiating and listening will always be important. If you ever move into a management position where you are accountable for others or a project, then these qualities will become even more important. Your ability to co-ordinate, inspire and lead will hinge on your people skills.

How to Regard Other People

I've said it before—first of all, get into the habit of treating other people with respect by default. If you allow the relationships that you have with other people to deteriorate, or fail to form new ones, then a big part of your career will be missing. Not only should you not be destroying your existing relationships, but you should be nurturing them! This is not as difficult or time-consuming as it seems.

Treat those closest to you particularly well. You may have heard the famous quote from Tsun Zso in *The Art of War*: "Treat your workers [i.e. your colleagues] like your sons and daughters and they will do anything for you". As stated in the previous section, you can rely on a natural justice in the engineering world in terms of your relationships with other people—in most cases, you will be treated according to the same standards that you establish with others.

Over time, you will notice that a small minority of people are treating you particularly well, and the reason is not apparent. If others appear to be helping you significantly, they seem dedicated to your cause, or value your every word, then my advice is to never turn your back on them! Cherish them and their viewpoints! There will be a limited number of people who will 'resonate' with you in this way throughout your career, and they may prove to be very important to you.

Realise that you are not at the centre of the world. Other people have their own priorities and you are rarely central to those priorities. They are living in their own worlds! It is too easy to imagine that it is all about you, and this leads to problems. This leads to us thinking that others are scheming against us when their decisions aren't favourable to us, whereas they are usually just thinking about themselves or their own interests. This knowledge also acts in your favour—figure out what they want, and try to give it to them. This sounds counterintuitive, but in order to achieve what you want with other people, you often need to give people what they want first!

Realise that all people are different—drastically so. They all have different goals and approaches. Don't let this frustrate and unsettle you, and don't try to change people. It may seem that your way is best, but you would be shocked to learn how effective other people can be, solving problems in their own unconventional ways. Just let them be and don't try to change them or their approaches (obviously I'm not talking about the cases where they are making blatant errors). *If you are a supervisor, then try to gauge people by the results that they deliver, not by the way they go about doing the job.*

It is also a good idea to give people the benefit of the doubt. It is amazing what our imaginations can conjure up. In particular we tend to make the worst of people. Working habits, approaches, frustrations, styles of communication and aspirations are different for everyone. People have their own reasons for their actions, and they rarely live up to our expectations. Of course, I'm not saying that you should tolerate inappropriate behaviour—you shouldn't! But realise that people have very good reasons for taking the actions that they take, and those reasons are hidden within them.

Get into the habit of praising people. All people love to be praised, whether directly or indirectly. Of course, I am not saying that you should do it excessively or offer praise when it is not warranted—this may come across as condescending and people will think that you want something from them! It is simply a good idea to give credit when credit is due. I don't think this is done nearly enough, but it is actually not as easy as it seems. Believe it or not, praising people while sounding (and being) authentic is a skill that you must develop. Some people will lap up praise no matter how you give it, but with most people you will arouse suspicion with praise that is not skilfully delivered, especially when you are not already known to praise people. Like most things, you will become good at it with practice.

Finally, don't let other people drag you down or put you off balance. Most importantly, don't allow anyone to unsettle your convictions and take you from the path you know is right for you. Do you have a good idea? Chances are that it caught your attention for a valid reason. As mentioned in Chapter 3, the way that people respond to you is always a reflection of the storm of emotions and memories of personal experiences raging within them. Don't take their negativity personally or let it affect you! If you were to take every criticism personally, your self-esteem would be at world-record low and you would be running around in circles, not achieving anything worthwhile.

Body Language

I mention body language in this book because it is useful to possess an awareness throughout your engineering career. While I only touch on the surface of the topic here, there are numerous sources available for an in-depth understanding. An understanding of body language may help you in understanding what other people are thinking and what their state of mind is. It can also allow you to more effectively deal with difficult situations. Furthermore, by becoming aware of what negative body language you are inadvertently giving off, you can aim to correct it. Needless to say, most of your interpretation of (and transmission of) body language will happen at the subconscious level; I'm sure you've heard the saying that "97% of all communication is achieved via body language".

You also have the power to influence people by faking your body language, but this is something that I don't recommend doing, since it is disingenuous and you can easily convey mixed messages (there will be a disparity between different aspects of your body language and what you say—this will trigger subconscious warning signals to others, and you will not seem congruent or trustworthy).

There are a few examples of body language that I will mention here. These are gestures that you can start to look out for, for practice. I mention these as interesting examples—not because they represent complete coverage of all possible gestures. There is far more to body language than this.

Of the three gestures that I see most commonly in the workplace, the first are the crossed arms. You will see this gesture very regularly. I see this mostly when I go to see someone for a favour. Often I won't even have had a chance to ask for the favour, but the person can sense what I am about to ask and so they cross their arms. This normally means that the person is defensive. They are not open to your words, and they are unlikely to welcome whatever it is you are about to say. I also see this gesture when I am offering advice, or trying to convince someone to go down a particular path, or buy a particular product, or do something different to what they were planning on doing. If you see this, then use it as a feedback mechanism to tell yourself that you have probably been a bit too bold, and need to back off slightly; you need to explain the problem in more detail or through a different perspective. Perhaps you need to relax some of your demands; adjust them make the task more palatable for the person. Crossed arms tell you that your sales technique wasn't quite right! Having said that, sometimes it is an unavoidable gesture depending on the mood of the person, and it certainly shouldn't be interpreted as a showstopper. In fact, in many cases the person will still give you the favour, but there may be some negotiation involved.

The second gesture that I commonly see is the hands touching the mouth or nose. This is a classic indicator that someone is lying. I rarely see this one with colleagues (luckily) but I see this every now and then with external people, such as clients. If you see a person making a bold statement and covering their mouth, it is highly likely that they are covering up something more than just a part of their body! The likelihood is magnified when they return for a second touch, or they then touch or scratch their nose. For example, once I had a software developer trying to sell me a software package. He was telling me that the software is extremely reliable and he has never had problems with it crashing. While telling me this, he was repeatedly covering his mouth and nose! Of course, he probably had no idea that he was doing it but it was a dead giveaway that the software had major problems. Later I found out that there were indeed countless bugs, and a bug list was created which had never been addressed.

The hand touching of the mouth or nose is a really simple gesture to spot, and once you become tuned in to it, you can spot it very quickly. I would advise against pointing out these behaviours to the person making them. Keep it to yourself and use the knowledge judiciously. If you point out a person's body language, then in all further interactions they will stop doing it and you will lose a valuable insight into their thinking. More importantly, their communication will become inconsistent (what I mean is that they will force a new pattern of body language upon themselves which may be inconsistent with their words or their subliminal body language, causing an incongruity in communication), so you would not be doing them any favours anyway. I'm sure you've talked with people like this, where there was something with their communication that wasn't quite right. They were probably trying to fake or force their body language after becoming aware of an undesirable quirk, and as a result will find it hard to build empathy with others. Don't point out other people's signals for your sake and for theirs.

The third gesture that I see regularly are the confident raised arms, with fingers crossed behind the head, often when sitting and leaning back in a chair. This is an indication of superiority and confidence. I normally see this with more senior people, and some younger tech-heads that probably know a lot more than me with whatever I have come to see them about. This gesture means that they feel superior; they believe they know more than you on this topic, and they don't feel threatened. I don't let this posture bother me one little bit. In fact, you can normally get what you want out of the conversation as long as you don't try to embarrass them or invalidate their authority.

CHAPTER 6

How to Continue to Make Progress

If you follow the advice in this book, then you will undoubtedly experience success—it is just a matter of time. Your level of success depends on exactly where you were at when you read it, how well you adsorbed the information and how effectively you applied it. However, what is more difficult than achieving short-term success is maintaining the momentum in the long term. You may have great success for six months or one to two years, but after that you will be tested, and you will have doubts. Your work circumstances may change; your family and financial situation may change. The good news is that no matter what your situation, in terms of your engineering career the principles in this book will hold true—you don't need to make any other changes!

Realise that the only reason why you would cease to make progress towards your goals is internally driven. That is, if you decide to give up! As long as the goal-plan-action machine is moving, then you will make progress. If you need more inspiration then reread this book, or find a mentor and role model to keep you moving. Perhaps there will come a time when you need to revisit your goals and find a new challenge. Perhaps you will need to spend time away (or time off) and change your surroundings and social groups for a while. This can be very helpful in finding new passions and new goals. Realise that there are plenty of activities out there that you would find compelling; it is just a matter of finding them along with the inspiration to pursue them. Very often it takes some-one else to steer you onto them and to inspire you. Sometimes they are goals that you just weren't ready for in the past, but you are now!

As I've said, by now you should have all the weapons in place to continue to make solid ground. Let's recap some of the more important ones:

- A vision and direction (motivation through a definite, emotionally compelling need, and a clear game plan— the vast majority of young engineers won't even make it past this dot point, and those that do will stop at the second!)
- No self-sabotaging behaviour (such as limiting beliefs and excuses)
- A balanced viewpoint (seeing things for what they are, and not for how you'd like to see them or how others see them)
- Massive action and drive (the single most important point; *without your deliberate action nothing will happen*)

- Perseverance and high self-esteem (not giving up and not letting others pull you down; virtually everyone including your closest friends and family will unintentionally do this)

Set Your Sights on Big Things

There is another reason why you may cease to make progress. Many engineers suffer this fate. They set goals that are not high enough. In other words, their horizons are limited. For example, consider the following career plan: I will get my degree and I will have a good job that will earn a living that is occasionally interesting. That's it. Then, after several years, they stop pushing because they have achieved their goal. The goal needs to be inspirational. It needs to be *beyond* your wildest dreams. And if you reach it, then make another goal and keep going! If your true goal is to have a cushy nine-to-five job, then this is fine, but do not be surprised if that is all you ever achieve!

You'd be surprised—in fact, you'd be shocked—to learn what you *can* achieve. I'll repeat this again ... what you *can* achieve is far greater than what you can imagine, but as I've said before, all great things must be worked at, and the results may come in unexpected ways! All great things are beyond your horizons. You know that they are there. You must have the faith to continue, believing that if you continue doing the right things, the spoils will come. So what are the spoils? The spoils are your wildest dreams—either the small ones or the grand ones. These are the things that matter to you, and only you. For example, your goals may be to work for a particular company, or earn $1M, or work in Bolivia, or lead a team of a hundred people, or start your own company! The big catch is time. Great things take great time and a lot of failures.

Staying Positive and Shaking Inner Demons

The further you go, and the more successful you become, the more people will (usually unintentionally) attempt to drag you down and destroy your self-esteem. However, you can't 'detach' from 'the system' either! I see many engineers do this. They will say it is not fair, or complain that others are harmful or the system is flawed. *If you lose faith in people and 'the system' then you might as well pack your bags!* You need to maintain an ongoing faith that people and the world and the system is good and they are in place to serve and help you.

You simply have to recognise that you will be constantly attacked and dragged down (in the vast majority of cases not intentionally and certainly not with hurtful intentions), and this is just the way it is. Nothing grand was ever easy. There are three sources of destructive, confidence-sapping put-downs:

1. Yourself. You are your own worst enemy, not only in your professional life but also your personal life. As engineers, we rarely give ourselves enough credit. We are generally modest people. We are constantly worried about not being smart enough, experienced enough, having the right background, being good-looking enough, being old enough, being too old and so on. It

is like a constant attack stemming from within. We are constantly sabotaging our chances of being successful in our careers. I would honestly say that 99% of the boundaries and limitations that stop people are introduced from within, by a faulty belief system. That is why it is generally better to overestimate your abilities, than to underestimate them. This sounds counterintuitive for the average engineer.

2. Loved ones and colleagues who have good intentions but can damage your career, easily setting it back ten years or more. There are various forms of this type of advice. Once you are aware of this, you will recognise them on an almost daily basis:

   a. Genuine, well-intentioned advice from loved ones such as family and very close friends. This type of advice generally discourages you from pursuing your goals. The reason is that your ambitions will rarely align with the ambitions of others or even your colleagues. Therefore, you will receive heart-felt pleas to pursue other paths or stick to the same, less-risky career path. The vast majority of damaging advice will come from this category. You may not even realise how damaging it is until years down the track. You have been warned!

   b. Bitterness and jealousy stemming from internal conflicts and battles experienced by just about everyone. I've mentioned before that people are complicated. There is little you can do in this case (though sometimes you may be able to distance yourself from the worst offenders), but it is important you recognise it!

3. Society, media and our 'culture'. We aren't aware of this since it appears as the norm to us, but the society around us instils a culture in us that can seriously limit us. Everyone has been led down a standard path for their lives, for example, go to university, land a job, work nine to five every day and so on. This does not mean that this is optimal for you. In fact, the greatest achievers are those that have torn away from what society called normal!

When realising that someone is dragging you down (whether intentionally or inadvertently), it is important not to engage them. You don't need to convince them how limited their horizons are, since from their perspective it is you who is misguided. It is also important to stop seeking their advice. By all means, take their advice on board, but if you continue to maintain regular contact with them, the risk is that eventually they will convey their limited horizons to you! It is therefore often a good idea not to broach the topic—try talking about different topics with these people. As they say, you are the product of the five people you spend the most time with—choose carefully!

Don't let the above imply that you should ignore everyone's opinions and do your own thing. You can also get into trouble this way. For example, quitting your job before you have a new income stream lined up to start your own business is generally not a good idea, and others may indeed be able to warn you of the perils. However, realise that others can also be major limiting factors to your progress. Don't underestimate how drastic their influence can be. The best approach is to get as many opinions as possible, and track down the people whose hearts are aligned in the same direction as yours! Better still, track down the people who have already accomplished your dreams. They will be able to explain the pitfalls to you with more credibility.

Ask

The saying 'ask and you will receive' applies to engineering and needs to be stressed to young engineers. I often see the extraordinary limitation that people impose on their progress by never actually asking for what they want. It truly amazes me!

Young engineers in particular seem to believe that all of the rewards (whether it is money, recognition, courses, conferences, travel, projects and so on) will just come along and lay themselves nicely on their laps, but this is rarely the case. By clearly asking for what you want, you will often get it, or part of it. At the very least you will be placed on a path to obtaining it. You need to be clear to yourself and others (particularly those above you) as to what it is you want.

Unless it is something illegal or unethical, then ask—even if you think the answer will be "no"; even if you will feel embarrassed for asking. Let's face it, you are not going to be around forever. You do not have the time to wait for things to happen themselves, and you can't assume that anyone else is aware of what you are after. If your boss says "no", then chances are your request will be quickly forgotten about anyway (obviously there are limits as to what you should ask—asking your boss for his wife's phone number will most likely not go down well, but asking your boss for a raise or to use the company car may be perfectly acceptable if framed correctly). I cannot tell you how many embarrassing requests I have made both in my professional (and personal) life.

As they say, if you don't ask then you don't get. It is easy to assume that others are on your wavelength and know exactly what you want, and that they will come to your rescue. Perhaps as a new recruit it is easy to believe this (and perhaps in the early days others really do tend to look out for you) but in the long run, this default safety mechanism will dwindle away. People have their own priorities and their own battles. Therefore, you need to make it clear what it is you want, where you are struggling and what help you need.

Believe it or not, asking also will build your confidence and people skills. Firstly, just getting your desires out there will make you feel better. Then there's the chance that your request will be met, or that other doors will open. Furthermore, by being able to frame often extravagant requests in palatable and diplomatic ways, you will become much better at it, and you will tend to put your priorities ahead of any social stigmas which can seriously limit your progress. Your people skills will also improve. Once you are able to ask for just about anything you like (within limits of course), you will be amazed at how much more open you will become! People will see you as someone who is transparent and trustworthy, and they will apply the same behaviours to you! You will also convey to others that you are a driven person and know what you want—very important characteristics in the workplace.

Don't Put People on Pedestals

It is generally not a good idea to put fellow engineers on pedestals. Don't misunderstand me here—you must still treat others respectfully, as I keep reiterating. However, assuming that others are smarter, more capable, less prone to errors and more driven than you is not a helpful belief to maintain.

Firstly, this is because it is simply not true. I have witnessed countless mistakes from the smartest, best-educated engineers that are hard to fathom. People develop strengths in various areas—even though someone might be a better presenter than you, they may have poor written skills, for example. Others have the same number of hours in each day and they develop different skills. Realise that you have strengths (or have the potential to develop strengths) that are very valuable to others. With time, I'm sure you will also spot many embarrassing blunders from other people, if you haven't already. Having said that, this law applies both ways. While you shouldn't place people on pedestals, you should also not believe that you are a superior being! I have seen this countless times from younger staff, particularly in the information technology area. They see someone doing or saying something stupid, and they immediately treat them like an inferior life form!

The real problem with putting people on pedestals is that you will tend to place their priorities and views ahead of yours. This is a fundamental blunder. Earlier in this book I explained the importance of role models and this view still holds. Of course, you need to draw from and model other people's successes. However, every person has his or her own goals and their unique way of going about business. Some of this can and should be replicated to fast-track your progress, but maintaining your uniqueness is also important. In particular, you cannot let others stamp out your visions and goals and unique take on the world. At the same time, the experience and ability of older, more established, or more highly skilled engineers can give you massive benefits if you learn from their unique abilities properly.

It is also much more fruitful to have a respectful, professional conversation with someone than to treat them like a rock star. From their perspective, they are normal. Grovelling and melting away in front of them will make them wonder what on earth you want and will make the interaction awkward. *People will respect you more when they see that you have respect in yourself, above all others.*

Be Strong

Up until now in the book, I have painted a picture of a well-behaved and respectful engineer—an employee who is malleable, always says "yes" and generally tries to please everyone. This has been a by-product of explaining the fundamental humble, appreciative and respectful behaviour that you should exhibit.

In order to make real progress, you will encounter obstacles, including other people and situations where you will have to say "no". These situations will challenge your values and plans and even your understanding of how the engineering world works! Some of these obstacles will try to stop you in your tracks. I find that most driven engineers are good at overcoming the material obstacles. For example, finding a new supplier, trying a new algorithm, implementing a new design and so on. However, other people can be a more difficult obstacle. This includes your peers, your supervisors, your subordinates, your friends and even your family when they have a say about your professional life. There will be situations where there will be no obvious solution, and no reasonable person is able to solve your problem. In cases like this, you must remain strong; recognise the complexity of other people and engineering, stick to your values and try to learn as much as possible while remaining steadfast.

Other people are a blessing and mean everything, but they can also be very dangerous, particularly if we are not clear about what we want ourselves. Since we are generally raised to listen to others and take their views seriously, and we naturally don't like confrontation, we tend to fold and bend according to others. Sometimes this prevents us from making a big mistake, but at other times this prevents us from making progress. *Virtually all of your truly big ideas will be opposed by others and all of the big ideas in history have been initially opposed by others.* At the very best they were faced with mild scepticism. It was the engineer's determination, resolve and vision (which meant putting their own opinions ahead of others) that allowed them to persevere. This can be very hard if the advice is coming from people that you respect.

I know I've talked about this previously, but it is worth revisiting this topic in this section on being strong. Other people will cause you to doubt yourself and to ask: is what I'm doing really worthwhile? Perhaps it is a mistake? If you are not strong, then you will succumb to their comments and views and it will stop you in your tracks.

In almost every case, the advice you receive from other people has good intentions. However, sometimes it is coming from the heart and sometimes it is their egos sabotaging them without their knowledge and trying to drag you down in the meantime. If you are doing something beyond their wildest dreams, or if they are set to lose from your action then you will be dragged down by small-minded people. Of course, they don't know that they are causing you damage; they are just going about their regular lives. *The problem is that other people are constantly and subtly imposing their limiting beliefs and values onto you, and normally you won't even realise.* The size of projects you pursue; the type of jobs you go for; the length of reports you write; the software you use; your work ethic and how many hours you devote; how you treat people around you and your behaviour in general: all of these things are potentially hampered by the influence of others. Of course, if you are small-minded yourself then you won't even notice. But the moment you start trying to break free and start thinking big, then you will notice.

So how do you know if you should take their advice or go your own way? The more people that you talk to, the better—especially those who have already achieved what you want to achieve. They will be able to warn you of the pitfalls and highlight the rewards. Write down the pros and cons on a piece of paper. Once you have done this, sometimes your mind just needs time being occupied with something unrelated for a while before it will automatically come up with a solution. When assessing someone else's advice, ask yourself whether it is bringing you closer to your work goals, or taking you further away. When their solution takes you in the opposite direction to where you would like to go, then red warning lights should be flashing!

You want to be firm and focused on achieving what it is you want to achieve. You want to resolve all conflicts in a diplomatic way and not burn relationships or bridges. Thank everyone for their input, but go about your business the way you think is best. This sounds simple but it requires a lot of strength, willpower and determination. This is where the strength comes in—not in arguing with others or having conflicts with others.

Leadership

When you become good in your branch of engineering, the best way to multiply your knowledge and your contributions is through effective leadership. This may not be apparent to you right now, but there may come a point when you realise that no matter how good you become at something, and no matter how hard you work, the impact that you are making is not as significant as you would like it to be. Becoming a leader is not for everyone, and I don't believe it should be forced upon anyone. It takes a lot of time and effort to lead. In addition to having the technical or scientific vision, it requires the development of people skills—specifically the ability to sell your ideas and inspire others, and the ability to deal with the usual people issues on a regular basis. I firmly believe that anyone can learn to lead, so if you think it is not for you then please have an open mind. Like any other aspect of engineering, if you are not enjoying it, then you are unlikely to push through and you will give up. It is also something that you may appreciate later in your career. There is no hurry, particularly if it does not appeal to you yet. Perhaps it never will and this is also perfectly fine. The important thing is not to have any limiting beliefs in this regard (for example, "I will never be a good leader. I don't have the people skills") and approach the possibility of being a leader with an open mind. Leading others can be an immensely enjoyable and rewarding experience for everyone.

Effective leadership plays a much more important role than you might initially imagine. By effectively leading other people, you are providing the critical focus and drive for the entire workplace. Therefore, your engineering prowess, whether good or bad, is magnified. Secondly, when you demonstrate how to be an effective leader to the people in your organisation, and you develop and nurture them as leaders, then they will begin to lead others too. Your efforts are passed on in two ways: in terms of your influence as a leader and your ability to shape future leaders. As a consequence, you can positively influence thousands of people throughout your career and beyond. Of course, if they leave the organisation then this effort is not wasted. They continue to be good leaders in other organisations and fields, and contribute to their new areas.

Contrary to what some young graduates might think, an organisation doesn't want all of the power in making decisions and leading its employees. This is extremely difficult! To lead a large organisation you must have excellent technical and professional skills; you must be up to date with your industry; you must maintain strong contacts, and an awareness of laws and regulations; you must have excellent people skills and be skilled in reporting and presenting, as well as the selling of ideas. Most difficult of all is the making of decisions—particularly for upper management. An organisation wants employees to stand on their own two feet. Most importantly, they want their leaders to steer the company in the right direction, largely autonomously. Good leaders are therefore critical to any organisation.

The downside of pursuing a leadership position is that you may not be able to carry out your normal engineering work. The best and most respected leaders in engineering often have absolutely no direct involvement in engineering! Of course, they will tell you that they keep in touch with technology and what is happening at the coalface, but the truth is that their time is much better spent leading others.

You must remember when pursuing a leadership position that, as a leader, your success will be defined in the same way it was as an engineer—for getting the job done. This is something that new graduates forget and this often causes them to lose perspective. At the end of the day, it doesn't matter what you know, who you know, who you are or how you did it. What matters is whether your abilities came together to get the job done (or contributed to getting the job done). A successful leader allows the people that they are leading to get results, to achieve milestones and hit company targets. This is the bottom line.

So what makes a successful leader? First of all, a leader can only lead others when they can effectively lead themselves. If you can't achieve results alone, then it will be difficult for you to lead others. This is because you are likely to lead them into the same dead-ends that you experienced as an individual! Secondly, it is difficult for someone unsuccessful to generate a substantial following! In other words, you must be good at what you do, and you must have a knack for getting the job done and delivering results. In a leadership position, it is as important as ever to have clear goals and determination. People will only follow you if you are driven and you know what you want and you have a clear path to getting it done—even if the path is not ideal. This is what this book has talked about all along! *Therefore, the progression from being a good engineer to a good leader is a very natural one.*

You must realise that almost all engineers (and others) are lost to some extent; they don't possess clarity of vision, and they are often plagued by workplace politics and inner demons. This is just the unfortunate reality of the human mind. As discussed in Chapters 3 and 4, people have egos and worries and exhibit harmful and self-sabotaging behaviour (people will often do the very opposite of what you have read in this book, and now that you know what to look out for, you will see it regularly!). Just about everyone would like a path made out for them that ensured them success and happiness. If you demonstrate your clarity of vision and your drive, and people can sense your imminent success and that you truly believe what you stand for, then they will be faced with a decision: either board the thundering train, or stay in the train station worrying when the right train will come along! In most cases, they will join you (and if they don't then they are probably not the people you want on your team anyway). *The best leaders have followers who follow them because they want to—not because of their status in the organisation or because they feel they have to.*

Even if you have no current desire to lead, you can learn a great deal about leadership just by observing your own leadership team. What do they do really well? What do they do that annoys you? The opportunity of being led is often just as valuable a learning experience as leading others!

Exercise 18: Think of three things that your current or previous leadership team do really well. Why were these actions effective? Conversely, think of three things that your current or previous leaders did poorly. What was poor about them?

As a final note, it is worth mentioning that your success as a leader is highly dependent on the team that you are leading. This is why it is so important that a company recruits the right individuals—in particular the right leaders! This is also why, as a leader, you should establish your vision and expectations early, giving those not aligned with your vision an opportunity to recognise this. Most aspiring leaders can lead a team of model employees who are able to set direction themselves. Inevitably, if you end up in a leadership position, you'll be leading a team of people who don't behave the way you may want—they may require significant input from you. Some people are just difficult to lead and this is where the role of a great leader really kicks in! This applies to just about any other area of engineering. For example, any designer can tackle an easy design problem—that is, something they enjoy and have done well before. But faced with a near-impossible problem, only a select few individuals will possess the ability to solve it.

The Importance of Improving

The single most important element in achieving long-term success in engineering (as well as just about every other career or aspect of life), is the ability to continually improve. Every day successful engineers become slightly better than they were the day before. People who go nowhere don't improve, or don't improve enough. They stagnate. They reach a point where they either become content with what they have, or they have no vision for further advancement. Often it takes a major event (such as a failed project, a failed subject, or a lost job) for them to muster the motivation to take action. This is one reason why failures are so important in achieving success. For this reason, the process of improving can be seen as one of two steps forward, and one step back.

If you are to succeed at engineering, you have to become good at engineering. It is really pretty simple! A healthy attitude is not enough—it is simply a tool for becoming a successful engineer. Regardless of whether you are a designer, or coder, or report writer, in sales or a manager, you will attain success by doing your job well. If you can do your job extraordinarily well, then this will attract the top salaries and open the most doors. But of course, every engineer knows inside that they must become good to be successful, but how do they go about it?

Firstly, becoming good at anything in depth takes time—perhaps ten years or more! This is something that makes most engineers uncomfortable. They think that if they can't succeed *right now*, then something is not right, but this is not normally the case. History is filled with examples highlighting how much time it takes to become really good at what you do (for example, starting a successful company or inventing a new product that makes an impact on the world), whether it is in engineering or any other profession. Learn to accept this and be patient.

The most inroads can be made by working to your strengths. What seems to come easily to you? What do you naturally enjoy doing? What will you keep on doing no matter what? These are the things that you are most likely to become great at. It is very hard to outperform someone who is doing what they love to do, and who just won't give up doing it—they just seem to do it well and they strive to become better and better at it, irrespective of what others say.

You should also think about the depth or the potential of the activities that you enjoy doing. Will they utilise your own full potential? For example, there is nothing wrong with designing broom handles, and it may be the most enjoyable experience for you in the world—however, it may lead to a career dead-end. You may reach mastery quickly, and you may realise you are not contributing to your full engineering potential to the world. If this is the case, you may consider broadening or shifting your scope to something with more depth that will still enthral you.

Exercise 19: Write down three things that you do really well. I don't mean things that you *want* to do well, but things that seem to come naturally for you. What successes did you have? Did it really seem like work? How have people praised you?

Beware of concentrating on aspects that you think you should remedy, or that others tell you you need to remedy. As I discussed above, work to your strengths and not to your weaknesses. Of course, I do believe that an engineer should be well-rounded, so I think you should do everything to a reasonable, professional level. In this light, if there is a glaring hole (for example, you struggle talking in public or don't have a basic idea of who's who in your organisation, or don't know about the basic ins and outs of your industry) then by all means fill in those glaring gaps. But I do not believe in spending years attending various training courses attempting to achieve mastery across a broad selection of topics—there are just too many possible topics and skills in engineering, and the more you learn about them, the more you realise how enormous these areas are and how little you really know! Very few employers pay top dollar for (or even appreciate) those engineers who are moderately good at all things. They generally pay top dollar for those who are the best (or close to the best) at one or two things. The other skills are simply a bonus for them.

It is easy to waste valuable time throughout your career pursuing skills that on the surface (at a particular time and place) seem important, but when looking at them in hindsight with consideration of your grand plans, really don't matter much. I have seen countess engineers waste their time pursuing common avenues they had limited interest in, yet were forcing themselves to tick off. Your career is too short to try to master skills that don't interest you. So unless you have to (because you are instructed to), try to focus on what interests you and what aligns with your grand goals. Even within the narrow regions of your interests, there are plenty of avenues that can keep you intrigued for many lifetimes!

I can't stress enough how important it is to rely on others who have already done what you want to do. They can pave a way forward for you and give you the motivation that you need. Let them pull you up to their experience level. Others are the key to you making gains quickly. For this reason you should also read! There is a wealth of knowledge out there from technical books to management books, communication books, books on work practices, mentoring books and so on. Don't be dismayed if a book doesn't solve it all, but even if you can extract one or two important points from a book then it has done its job. Education is one of the keys to improvement, and I'm not just talking about the technical side. There is so much wisdom out there, but it is often hard to find the information that you need. Set up a reading list and update it regularly. You will hear of new books, journals or online resources through word of mouth or reviews. Try to gauge what each source is good for (as I said, normally it will only bring you a limited benefit, so you need many sources). I am constantly working my way through a resource list—it is never-ending!

In addition to developing new skills, try to raise your standards in what you can already do. It is amazing how much better we can all become at the skills that we are already good at. This is because once we become the resident expert, we feel relatively little pressure to develop further. For example, if you are passionate about a particular software package, consider subscribing to the relevant email bulletins or attending a sales talk; try actively getting involved in the community; become aware of the updates and bug-fixes; explore the tips and tricks of the package; develop an awareness of who else is good at using it; enter competitions and challenges; challenge yourself with the software (for example, to create a program with fewer lines, or learn a new library).

Realise that if you are working with technology, the learning never ends! Every kind of engineering has depth and complexity that is beyond any individual's comprehension, and it is quickly evolving. In fact, if you do have one enemy in your career it is time! It may not seem like it now, but once you start developing different aspects of your career, you will realise just how precious time is. Time seems plentiful when you are in your twenties, but this is not the case once you have established direction and a solid game plan.

Finally, do not stress about going down a 'learning dead-end'. For example, if you invested several years of your professional life learning about accounting for engineering, only to realise that it is just not relevant for you, then have faith that you chose it for a reason and that this knowledge will eventually come in handy. What is important is that you took action and learned something new. This is why you should think twice about attempting to 'undo' a choice that you made, such as quitting a degree (though in some cases this does indeed make sense). I've heard the saying that life is like a river, and that you should just 'go with the flow'. I think that this saying applies very well to engineering. You can control only the general direction that you go, but not the exact path. Fighting against the current at every bend will only wear you out, wasting your time and energy. Set the big goals and make the important decisions, but allow the time and space for everything in between to flow and pan out.

Realise that it is easy to work on your career when you absolutely have to. It is another thing to work on your career when you don't have to—when you are mostly satisfied with it. Most people who are mostly satisfied with something do not make an effort to change it. However, in order to go places and be an outstanding engineer, you must be constantly working on your career. The difficulty is conjuring up the motivation to do so. This is why it is so important to have those ambitious goals and compelling reasons worked out—they will force you to take action! Not taking action will simply become too painful for you to bear!

This is one of those important areas where you can differentiate yourself from the rest of the pack—improving at times when others don't, when others are satisfied. Make no mistake about it—this is the toughest stage when it comes to continued progress. Once you have actually achieved something and begin to feel accomplished, you risk becoming content and fearful of losing what you have achieved.

In my opinion, there are six key ingredients required in order to continue to make progress:

1. Set your goals and standards high in the first place! If you set your standards high, then you will not tolerate anything less than you think you deserve and you will rarely be content! Perhaps it is time to revisit your plans that you made earlier in this book.
2. Use the goal setting you learnt in the earlier sections to break tasks up.
3. Spend time focusing on your goals daily. If you do not focus on the prize and want it badly, then it is likely that

you will lose motivation. You need to have a clear and compelling reason to persevere.

4. Make regular and honest assessments of your progress. Are you moving in the right direction? Are you moving at the right speed? Do not make any excuses when doing this.

5. Seek out ideas and advice. Talk to others about your problems. How can it be solved? Think outside of the box.

6. Do not be afraid to change direction as required. The most successful projects require hundreds of adjustments. A fresh avenue is often all you need.

Beware of Misleading Advice

Throughout this book I have placed an emphasis on learning from other people to save time. The negative aspect of learning solely from others is that you risk taking flawed advice. Yes, you can receive flawed advice and it is very common. The reason is not as obvious as it may appear.

Sure, people may not know any better and dish out harmful advice—but in most cases it is simply because people don't see the world from your viewpoint and don't necessarily know what is best for you. I have said this before but it is worth re-iterating. They don't see your values and priorities and don't understand your drive and determination (if it differs to theirs, and it usually does). Therefore, the advice that is good for them can be catastrophic for you.

Finally, people are complex and have their own egos and values that can sabotage you. Generally speaking, judge each person on the merits of their advice. However, you should also consider their achievement in the field you are seeking advice on. If they have achieved what you want to achieve, then this may be a good indication of their potential to offer good advice (though I've experienced many cases of receiving terrible advice from successful people so treat them as a good starting point only). Conversely, if they have failed at what you would like to achieve, then they are probably not the right person to be asking.

Also, as I've pointed out before, beware of taking advice from family and close relatives or friends who mean well. There is always the additional potential for unintentional sabotaging self-interest when personal relationships are involved. Furthermore, the closer the person is to you, the more they are invested in you and the greater the strength and determination of their advice! This is fine if it is good advice, but very often it is far from perfect.

Finally, beware of people's egos and viewpoints. Often it has taken me several years to learn why someone's advice was not quite right or tainted with some self-interest. Like all things, you will get better the more you practise. Therefore, get used to seeking out advice and getting other people's viewpoints. You will learn to read between the lines and understand where people are coming from. You will then understand the true foundations of other people's opinions, and how heavily coloured they are, and how they must be interpreted with care, even though their opinions will still be valuable. In fact, in just about every case you will be able to derive some value when you seek advice from someone, particularly if they are more established and experienced than you, but rarely will you receive the complete answer.

Adapting

It is important to realise that your projects, organisation and career will undergo change and this is healthy and normal. Yet it never ceases to amaze me how many engineers are opposed to change (there's always one on the team). They believe that their professional lives should be straight lines, and any perturbations should be avoided at all costs. In fact, the most outstanding and successful careers often seem to have the most bumps and bends! For example, your workplace may be restructured from time to time, your company is sold, you land new bosses and subordinates, you become involved with new projects and so on. Sometimes your work priorities will change completely or your entire engineering career may swing around. You may be forced to change jobs or even disciplines! It is important not to become dismayed, believing that change represents failure on your part or your organisation's part. Healthy engineering means a constant state of change.

Knowing this, you should be open to change. Accept and greet change with a smile! Too often I work with people who are so stuck in their ways that whenever a proposal is made, they immediately oppose it. They believe that practices and people should always remain unchanged. However, a company must evolve to meet with the ever-changing needs of the customer and client, as well as with rapidly evolving technology. Sometimes a company makes mistakes and needs to readjust, or it needs to improve its efficiency (it should always be seeking to improve its efficiency). Interest areas and demands change, technology areas change, staff skill levels change and new business approaches come along. For all of these reasons, a company needs to be making constant adaptations and innovations in order to prosper and survive.

Any technology company in particular needs to be very light on its feet, with the minimum of administrative and bureaucratic overhead burdens to tie it down. However, this philosophy does not apply only to high-tech companies. Even large and established engineering firms making a steady business from, say, selling steel need to constantly adapt or they will eventually suffer by stagnating, not achieving the profits that they are capable of achieving, or collapsing altogether! *Being prepared to make changes and implementing small changes constantly makes a company and its employees more capable of dealing with bigger change, which may be eventually forced upon it!* When you hear of a change presented to you, be curious and inquisitive and welcoming. Your default attitude to change should be positive. When a company or product doesn't change for years on end, then something is probably wrong! That company will not be able to adapt when something goes wrong or when change is really needed. There should be a constant evolution and feedback cycle at play for every engineering and technology company.

Change, of course, leads to uncertainty in the workplace which makes most people uncomfortable. However, it should be apparent to you by this point in the book that uncertainty is a normal ingredient of your engineering work and career. Uncertainty can give you motivation and vision to pursue new ideas. The world would be a very different place without it. Uncertainty and change lead to challenges and opportunities. If you have read this book carefully from the start, then you should know that these are the very ingredients that should excite all engineers!

Work-Life Balance

The engineers who are truly successful often seem to have many other aspects of their lives worked out. When I say "worked out", I don't mean that they are great at everything. This would be impossible. They may be great at several things, but the other aspects of their life are under control. This is largely because the same determination, drive and vision they possess in one part of their lives translates to others.

Other elements of your life may include:

- Your relationships
- Your finances
- Your living arrangements
- Your physical health
- Your mental health
- Your spiritual health

Successful engineers may not be experts in all other areas of life. Anyone who is truly dedicated to their careers won't be able to put the same level of commitment into the other areas of their lives, and they won't be successful across everything. However, they are constantly chipping away at the other areas.

One reason why it is worthwhile focusing on other areas of life is that it gives you balance. If you are happy across the different parts of your life, then you will become better and better at what means most to you. Beware, it is unlikely that you will fulfil all your needs from work alone! According to Tony Robbins, there are six fundamental human needs:

- Certainty and comfort
- Variety
- Significance
- Love and connection
- Growth
- Contribution

Some of these may be met by your workplace, but it is highly unlikely that all of them will be. For example, in one engineering workplace you may achieve variety, significance, growth and contribution. However, it will be up to you to fulfil the other needs using other means and settings.

Furthermore, you are likely to have other responsibilities anyway, for example, family, hobbies, sports, social circles and so on. It is also perfectly normal to spend several years concentrating on one aspect (such as growth, for example), before switching to something else. Since our life needs, interests and priorities change, different areas take precedence over others naturally. To become good across everything takes decades, but you have plenty of time. While this book does not discuss other areas in any detail, it aims to make you aware of them.

Succeeding in Other Areas of Life

In general, it is a good idea (whether in your professional career or in your personal life) that you explore some topics in depth, and have a passing interest in others. This is what naturally is likely to happen anyway. A good rule of thumb is to have five areas that you are really good at. For example, you may have two areas of expert knowledge at work, as well as one or two hobbies or sports, and one or two other extracurricular activities (such as tutoring or blogging or even video games!). American colleges traditionally looked for a well-rounded student who could demonstrate depth in a few areas (the 'comb' model of skills), which I think is a good model to go by, as much as possible (sometimes different parts of your life will impinge on others so this will not always be possible). This is also just a guideline. For example, you could be really good at several things, or perhaps ten things.

Achieving this sort of balance is not as difficult as it seems. The same principles that you have used to become successful at engineering can be used to become successful in other areas. You may see this in your role models. For example, they may have a great career, but they will also have a loving wife and a nice house, be in good shape and possess a peace of mind that others can only envy. It is too easy to think that they have it all, or that they were lucky, or that they were given it all or part of it. *The truth is that they probably worked very hard towards one aspect of their lives, and then realised that the same principles of success can be used to achieve success in other areas of their lives.* Once you commit yourself to becoming great at engineering, or perhaps some other activity or hobby (through clear focus, hard work, perseverance, seeking help from others, constantly learning and the other principles I discuss in this book) then it becomes relatively easy to apply the same principles to the other areas of your life.

Patience is the often forgotten element here—it will take a massive amount of time to achieve success in all of these areas. These other areas won't magically fall into place once you have mastered one aspect. Each one requires focused effort. Certain elements will come more easily to you than others, but you must realise and truly believe that no-one 'has it made'. For example, relationships with people may come easier to some people, but perhaps they may struggle with technical details, or perhaps they are not as good at maintaining and nurturing those relationships.

Successful people leave the impression that they have done it easily, that they were blessed with what they have, that they are naturals, or simply that they got lucky. However, this is just about always far from the truth. Very few people will admit to the amount of hard work they have put in, and the difficulties they've had along the way: broken relationships and divorces, unpaid bills, frustrating jobs and so on. Unfortunately, these are realities of life that successful people have to overcome to achieve and maintain success!

The media collectively has particularly unhealthy and damaging views when it comes to achieving success. The media tends to divide people into the 'haves' and 'have-nots'—it gives us the impression that success is created by 'good genes', or luck, or given to us as a gift, and often comes at the expense of others, but all of this is almost never the case! The media likes to create an impression of a talented and gifted individual who can succeed with minimal effort (i.e. a superhero) since this notion is appealing to most viewers. Few people want to hear about other people's successes brought about by hard work. They either want to hear about the failures or the successes that can be attributed to luck, or money, or family, or genes but these are usually minuscule factors when it comes to success. Viewers then have a convenient excuse (at least internally) for not achieving the same success and feel good about themselves.

One of the reasons why a work-life balance is so important is because all of these areas are required for you to be operating optimally in any one area. For example, if you are successful at work but in very bad health (because you don't commit any time to your diet or fitness), then you will not push the boundaries at work, since you won't have the energy or the mental freshness. Worst of all, you won't enjoy life as much!

This book does not delve into addressing each of these areas, other than explaining that they are intrinsically linked to your work. You should consult the appropriate experts and read the right books and websites. For example, if you want to address your diet or fitness, seek the help of a dietitian or trainer, jump onto YouTube, or join a gym. There is so much valuable information out there and so may options.

Become active in the communities that interest you. Seek out those interests that for some inexplicable reason take your fancy. What kind of sports do you like? What have you always wanted to do? What do you seem to be naturally good at? Has anyone ever complimented you on something that you did well? What does your body seem to respond well to? Do you like running or does it make you sick? What makes you happy?

My summary of this section is simply that your career alone will not make you happy and fulfilled in the long run. We have many needs that have to be addressed eventually. Luckily the world is a big place with plentiful areas of interest, challenges and rewards which are bound to take your fancy along the way.

Avoiding Stagnation

Every engineer knows what stagnation is. In a profession that demands continuous learning, hitting a roadblock can be a miserable and demoralising experience. It happens to just about every engineer at some point. It can leave you feeling like your career is headed nowhere, like you have stopped learning and moving forward, like you are no longer giving your best and solving the types of problems that you'd like to be solving.

The good news is that there is nothing new that you have to do to avoid stagnation; everything you need to know has already been explained in this book! The process that I've explained in this book simply needs to be reapplied over and over again. There is no end! However, you must be *actively pursuing the principles in this book, and allowing a healthy dose of self-reflection which must be ongoing.* I often see stagnation in resumes, where people seem to get derailed, but the derailment isn't sudden; rather, it is a slow process that goes unnoticed and uncorrected! Engineers generally don't stop to think and reassess whether their careers are heading in the right direction.

Exercise 20: Prepare a resume for yourself ten or twenty years down the track. Are you happy with it? What would you change? Change it until you are happy!

While your behaviours should be the same in all stages of your career, there are some changes that you can expect when you move further along in your career. It is important that you don't let these changes shake you. As your career progresses, you will generally:

- Feel like you have more and more flexibility and control with your decision making, and the decisions asked of you are more difficult to make
- Feel like the world has become a bigger place. You will realise how little you know and how small your piece of the world really is. You are not at the centre of the world as you once thought
- Realise that you have the power within you to change just about anything and to do just about anything
- Notice that direction is given to you less often. You have to find your own way. You are constantly at a crossroad and while other people will try to influence your decisions, the turns you make are ultimately up to you
- Notice that you are getting less and less attention for just being you, and more and more attention for your actions (what you did)
- Notice the massive differences in people's viewpoints and approaches. You can't let these shake you or change

your own way of thinking (unless you are obviously wrong)

These changes will mean that you will appreciate and pay more attention to certain parts of this book. Although all of the tools are already in this book, I have highlighted the ones that are particularly important in this section.

First and foremost, you will notice that you have to be proactive, much more than you used to be. As the attention focuses off you, and as the world seems to expand, *you will need to make things happen rather than wait for things to happen* (which was a reasonable expectation as a student or young graduate). No engineer has achieved great things by waiting for things to happen. I am continually amazed at mid-career engineers and even some senior engineers who are expecting management to come up with a path for them, in terms of approaches, the solutions, marketing and so on. And when this doesn't happen, they feel let down. *As your career advances, what you achieve will be up to you and you have to make it happen.*

Realise that you have a long way to go in your development—professionally and personally. The sooner you realise this and act, the higher quality positions you will hold, the better services you will provide, the more money you will earn (if this is what you want) and the higher quality relationships you will hold. When you finish school, you may feel on top of the world—like you know it all. After university, you begin to realise how little you know and how much you still have to learn. However, twenty years into your career, you will see the monstrous mountains that lie ahead in every aspect of your professional life: communication, leadership, technical skills, salesmanship and so on.

So how can you continue to make ground every day over the coming years? My first tip is to do your job, and do it well! This is the golden rule of being an engineer. When all is said and done, when all the fundamental techniques are discussed, at the end of the day all that really matters is how productive you are. This should be your number one long-term strategy.

Don't procrastinate or put things off. When I work with new starters, one of the most striking characteristics I notice is their willingness to put the difficult tasks off for later. This is a lesson that is painfully learnt. The problem with this approach is that it doesn't solve the task—you will eventually still have to do it. It will use up valuable resources in your mind and weigh you down, and it won't be fresh in your mind when you decide to tackle it later on (plus you may have to drop what you are doing). The cost of completing the task will generally be far greater when you put it off! Experienced engineers will do it immediately and won't put it off, particularly if it is something hard or uncomfortable to do. They will do it immediately even if doing it immediately means that it won't be done to the same standard. It is too easy to make a habit of procrastination. It is way too easy to substitute the important tasks with something easier that wastes our time and doesn't achieve much in the long term.

Play the odds. At the end of the day or month or year, everything that happened in your professional life will be perfectly justified. You may not know it now, but there will be few surprises when you look over your professional life later down the track. Therefore, you should play the odds—do the things that are most important to you, and are the most likely to get you where you want to go. It is disconcerting to be told as a new engineer fresh out of school that your professional life is basically a statistics class. However, numerous successful people from numerous fields even outside of engineering (such as investing, business, marketing, not to mention sport) usually achieve their success by observing the odds and playing the odds. Do what is proven to work, but do not interpret this as being safe or unoriginal. Furthermore, do not let emotion make more (or less) of a situation. See things at face value, and practise doing so regularly.

Continue to trust your instincts—they will serve you more and more, and you will learn how to utilise them more effectively. This doesn't only apply to the direction you take and the projects you take on, but the people you work and build relationships with. We often know in our gut whether something is right or wrong, and this feeling can be a handy guide. With time, your instinct should be harnessed and exploited to its full potential. This is possible because of the fact that most processing is done by the subconscious part of the brain. As your day goes by, there are millions of inputs that you are not even aware of, all processed by your brain. And it is not the conscious part of the brain that does the majority of the processing—it is the subconscious part. Since it is the subconscious part of the brain learning passively, then its output is also passive, in the form of 'gut feelings'. When people say "trust your heart", they don't actually mean your heart—they mean this part of the brain that we are not even aware we have. This part of the brain will continue to serve you and you will learn to listen to it, judiciously.

Continue to grow while being patient. All great things take time. A human life from the time it is conceived grows at a rate of about two millimetres per day, at its peak rate. This is the same approach you should have towards professional growth—every day you should grow a tiny amount so you are slightly better than the day before. Growth never occurs immediately (it just seems like it does since the rewards can come very quickly). It is also timely to remind you to continue to learn from your mistakes. This means being forgiving of yourself, but also realising that others will make mistakes so you must be forgiving of others too!

Continue to take responsibility for your work and your career. I've said this before and perhaps you've applied it once and seen good results, but this philosophy mustn't stop. *Realise that everything positive that happens is because of you. Realise that everything negative that happens is because of you.* Others have nothing to do with it. It is your actions that determines your results. Please reread this! Do not use the word 'luck'. As previously outlined, successful people in general believe that everything that happens to them is a result of their actions, even if it is not apparent. Unsuccessful people put it all down to 'luck' or 'bad luck', or other people and external events. Take full responsibility and don't expect to get anything from anyone. If you would like a raise or promotion, or you would like to work in a new area, then take action and make it a reality.

Let your hair down and indulge occasionally. You need time to relax and let your mind wander and solve other problems.

Allow unstructured time when you just chill out. If you are constantly tied up with thoughts relating to your work, then you are unlikely to experience good ideas. The best ideas come when you are relaxed—occupied with something else. For example, go hiking, play a sport, play a video game, watch a movie, spend time with family. When you are doing these things, you should be isolated from your work. As I've said earlier in this book, I believe that 90% of your time you should be on the ball, acting consistently with your goals, but 10% of the time you can just relax! After all, life is there to be enjoyed, with friends and family, or even alone. Though I believe your work mission is important, if you don't spend a reasonable fraction of your life enjoying it to the fullest and developing meaningful relationships with those you love, then eventually you will feel like you've missed out on life!

Keep it simple and don't overthink things. Have faith that if you develop great clarity of the problem, then your mind will find a solution. Think about things in a simple fashion. *The greatest minds and academics have this innate ability to take complex concepts and shrink them into something embarrassingly simple.* The people who aren't clear thinkers, or worry about their academic ability, or wish to prove their ability, or have strong egos all do the exact opposite—they take simple problems and concepts, and blow them up into complex concepts and problems, using unnecessarily complicated language, jargon terms and acronyms, and mix in irrelevant factors. This not only complicates things for others, but it prevents you from thinking clearly yourself! It also gives a 'reward' to your ego seeking an intellectual showcase for essentially doing nothing useful. Reduce problems to their smallest common denominator and you will have the capacity to tackle even bigger problems; you will be less worn out and you'll stand a better chance of solving them!

Remain positive and try to maintain high energy and enthusiasm every single day (or at least for part of every single day). You have all your wildest dreams to live for! No-one can do this all the time of course. If you are working at what you enjoy and what you would like to succeed in, and you are making progress by following the principles in this book, then you will naturally become positive. This will give you the energy to leap out of bed most mornings, knowing that you are on the right path (that is, your path) and slowly succeeding.

Realise also that you have flaws. Big ones. Everyone has. No-one has the perfect background or has made perfect decisions. Virtually all of the most successful people that I have met have become successful because of the way that they dealt with some massive problems, and not because of a lack of problems! Far from it! Even if you did make the optimal decisions based on the information that you had, we rarely have all of the information that we require! As I've said before, we are looking through coloured lenses. We think we are at the centre of the universe but this is not the case. The world is not really how it seems—it is very different. On top of this, our priorities change and the optimal decisions yesterday are not necessarily the optimal decisions today. Furthermore, it is often your dissatisfaction with your flaws that can cause you grief, rather than the flaws themselves. If you are worried about something that has happened in the past, then learn to accept the situation for what it is—it could not have been any other way! If you are dissatisfied with your abilities and current situation, then this is something that can usually be changed and should warrant your full attention.

You will eventually get to a point where you are mostly satisfied. Perhaps it is through following the advice in this book, or perhaps you would have gotten there anyway. I don't want to deflate your pride and put the brakes on your momentum, but everyone reaches this point! M*ost people stop progressing because they feel satisfied.* It is not the best job in the world; it is not the best situation; it is not the best pay, but it is good enough. This is why it is so important to have grand goals!

Realise that any progress in the absence of ideas is better that no progress at all. If you are wondering about completing a qualification, and you have nothing better to do, and no-one can you advise you otherwise, then if the alternative means doing nothing, then complete the qualification and don't look back! Reassess your situation when you are done. Making progress even if it is not in the ideal direction is infinitely better than not making any progress at all. It is only when this loss of direction is a habit (due to a lack of values, priorities and plans) that this can become a serious problem.

Also, be aware that there is no 'normal' career path, other than what society and the media tell you! Don't ever be discouraged that you have not progressed along some tried and tested path. In fact, it can be a good thing that you didn't! Some of the most impressive engineers that I have met did not take the standard path to achieve their success.

Do not underestimate how much time and effort it takes to achieve what you really want. Most people will just give up along the way. Virtually all of the really successful people that I have read about have said that their achievements took ten times longer than what they initially thought they would.

You know that your career is working for you when you are hitting your milestones, yet you seem to be on 'autopilot'. That is, you have a clear and distant view of where you would like to be, you have a set of goals, and most importantly you are making some progress every day. You know this because you have measurable milestones. You gain satisfaction from learning every day—learning about yourself, learning about other people, and improving your skills in engineering. You endeavour to understand yourself—what your weaknesses are, what frustrates you, what your strengths are and what gives you satisfaction. When I say that you are running on 'autopilot', I don't mean that the months and years are ticking away with no change—far from it! What I mean is that when you are busily achieving, then progress should seem effortless—like steadily sawing through the trunk of a tree! Of course, you are always experiencing the peaks and troughs. We are powerful creatures when we have a clear aim since we can largely 'switch off' and allow the subconscious mind to execute our aspirations.

Finally, as I've said earlier in this book, I have heard the analogy before of your working life being like a river, and how you should float like a leaf and not try to fight against the current. There is significant value with this philosophy, since you definitely don't want to go against the grain and spoil your good work to date or run around in circles since you will wear yourself out. It is too easy to think it is necessary to 'un-do' something that was perfectly normal. The problem is that sometimes when events happen and it is not clear to us what function they served (even when they are brought about by our own actions), we tend to question them and doubt ourselves. With time, the reasons for these events usually become clear so we just need some patience. The counter argument to this philosophy is that we need to be driven and form our own destinies—a valid point as well. Admittedly, there is a fine line between going with the flow and overcorrecting. Sometimes you have to allow for your career to flow effortlessly, while at other times you need to grab the stern with both hands—there is no firm rule. This is another one of those skills that become stronger with experience and through the application of the principles in this book.

CHAPTER 7

Congratulations

You now have the fundamental ingredients for succeeding as an engineer. You understand your values and goals and you have a plan in place. You also understand what constitutes a healthy inner attitude, and you know how to behave around other people. You have a bagful of knowledge at your disposal for achieving your goals in engineering!

While applying what you have learnt will give you some immediate results, it is the application of these principles in the long term that is most important. These principles must become habits, and the best way to form habits is to repeat the behaviour again and again until it sticks. You may consider reflecting for a while and rereading this book several months down the track.

If you feel like you have learnt three important lessons throughout this book, then you should be happy about your progress. I rarely find that any career guidebook offers me more than three important findings, so don't be upset if you found that much of this material was obvious, or if you didn't agree with some of it. It takes a lot of reading and many experiences to find the information that you need that is specific to your needs (though I have obviously tried to place what I consider to be my most important findings in this book). Along your journey in engineering you will pick up more information through your reading and interaction with other people. You will need continuous input from books, mentors, other contacts and experiences. There is rarely a silver bullet for anything! Like you, I am constantly learning new tricks every day and I'm moving forward every day.

However, I am hoping that you have learnt more than three things, and that this book will have a profound effect on your behaviour and your career. Hopefully, it will save you many years of wasted time and effort, as I know it would have saved me! In fact, I am hoping that you are now on the path to achieving major successes in engineering. Most importantly, just the fact that you have read this book means that you have taken action in the right direction, which already positions you in the top few percent of new starters. The seeds have been sown for a massive tree to flourish! But beware—you will notice over the coming months that most of your colleagues do the very opposite of what this book says! They have no plan, they make excuses, and they assume that other people and their organisations are the enemy! No wonder most engineers are stuck in a rut!

You are now in a position to bring together all of your knowledge, and combine it with some healthy youthful drive, clear direction, resilience and integrity, and the possibilities for you are far greater than you could imagine. But you have to make it happen—you have to roll the red carpet out for yourself and march along it proudly. You do not want to be standing on the sidelines watching, or you will regret it later.

As a final note, I'd like to restress the message from some of the great thinkers, philosophers and achievers—that life is fragile and short. Don't put things off—work towards what motivates you and what 'your calling' is. Respect the things that are closest to your heart—whether it is people, technologies, attitudes, or approaches. You will only regret turning your back on them! Don't become discouraged because certain things are hard for you or because you do things differently or that you seem to fail more than others. This is normal—it was meant to be and could not be any other way! Think of yourself as lucky to be here, and lucky to be in engineering!

Congratulations on making it through the book. I wish you every success in engineering that you can imagine!